MEMS-Based
Integrated Navigation

For a listing of recent titles in the *Artech House GNSS Technology and Application Series*, please turn to the back of this book.

MEMS-Based Integrated Navigation

Priyanka Aggarwal
Zainab Syed
Aboelmagd Noureldin
Naser El-Sheimy

ARTECH HOUSE

BOSTON | LONDON
artechhouse.com

Library of Congress Cataloging-in-Publication Data
A catalog record for this book is available from the U.S. Library of Congress.

British Library Cataloguing in Publication Data
A catalogue record for this book is available from the British Library.

ISBN-13: 978-1-60807-043-5

Cover design by Vicki Kane

© 2010 ARTECH HOUSE
685 Canton Street
Norwood, MA 02062

10 9 8 7 6 5 4 3 2 1

Contents

Preface

Microelectromechanical systems (MEMS) are an emerging technology that has the potential for a multitude of uses. The sensors are low cost, low power consuming, and easily available. MEMS inertial sensors have introduced a new concept of affordability to inertial positioning and navigation, which have always been dedicated for special applications due to the extremely high system cost and a very bulky system design. The new MEMS inertial navigation is unique and challenging in different aspects, but is readily available in today's market. This low-cost navigation technology has remained a secret for a long time, and little is known about it in the public domain that created a void in the well-established literature of navigation. As a result, there is a need to cover the new MEMS-based inertial navigation. Throughout the book, the authors aim to provide a balance between the theory that is mostly unique to this type of navigation and the actual field results to fill the existing void in the current literature. The book also covers very special topics that are only relevant to MEMS-based inertial navigation, which are of great importance to the commercial sector. These topics include minimizing costs by using different forms of aiding or removing inertial sensors.

Dr. Priyanka Aggarwal and I discovered the absence of comprehensive reference for MEMS-based inertial navigation during our time as Ph.D. candidates with the Mobile Multi-Sensors (MMSS) research group at the University of Calgary. Founded and led by Dr. Naser El-Sheimy, the MMSS research group is an internationally known cutting-edge research group in the field of navigation and photogrammetry using very low-cost MEMS-based sensors. Like the rest of the authors, we were also involved in publications, and at every publication we faced the same challenge of finding at least one textbook that covers some of the specialized MEMS-based navigation topics. Our problem went from bad to worse when we were asked by the referees to find journal references that were not produced by our own research group. There came a time that one of our joint publications was rejected because we couldn't find strong references for MEMS-based inertial navigation in terms of books or journal papers. That became the turning point.

Under extreme frustration, we decided that we need a book on this topic. The next stage was to find a comprehensive list of such topics. Our first thought was to take Dr. El-Sheimy's notes and convert them into a book; those notes used to be our source of general reference. However, as we started this journey, we quickly realized that we had to move our focus beyond the course notes and include topics that were either too difficult to find, too difficult to comprehend, or just simply not there. During this process, we also realized the importance of having examples from real field tests. To cut a long story short, almost 8 months after working on this book, we had all the unique materials for the book, but it was random in nature. Fortunately, by this time it was December of 2008 and I was done with my Ph.D. defense. That left me with almost one month for writing before the commencement of my new job with Trusted Positioning Inc. I spent most of December on the most important part of writing this book, which was arranging every chapter and working on the chapters' flow as per recommendations from Dr. El-Sheimy and Dr. Aboelmagd Noureldin.

The book that we have today is a combined effort of all of the authors. Dr. Noureldin had the difficult task of going through the material, editing it, and ensuring that it is theoretically and logically sound. Additionally, Dr. Noureldin wrote Chapter 7 "Artificial Neural Networks," to cover artificial intelligence techniques for MEMS-based navigation. Dr. El-Sheimy provided us with the valuable advice from his unsurpassed experience in MEMS-based inertial navigation in addition to supplying us with his course notes that we incorporated in different chapters. Chapters 1, 3, and 8 were written by Dr. Aggarwal. As for Chapter 2, it contains material from Dr. El-Sheimy's notes and both of us compiled and incorporated additional sections to it. I wrote Chapters 4 through 6 with editing work by Dr. Noureldin. As Priyanka and I moved on to different paths, it became increasingly difficult to combine our resources, and this introduced the need for someone who could help in the book formatting, front matter, and so forth. It was a tedious task and I found the help from my loving husband, Najmy.

We have made every effort to write the book in a way that enables the reader to establish the knowledge and expertise of calibrating and processing MEMS inertial sensors, modeling their measurement errors, and integrating them with GPS. This book requires a reader with enough knowledge and understanding of integrated navigation systems to be able implement the algorithms and methods provided herein.

It would not be possible to complete this text without the help of many people, whom we thank greatly. Ms. Heather Joyce Maki went through the whole book to ensure that it was grammatically sound. We realize that this was not an easy task but she did it for us with so much devotion that we are indebted for her tremendous favor. A special thanks also goes to Namrata Khemka for helping Dr. Aggarwal in reviewing the book. Among the teachers and mentors, Dr. Aggarwal and I acknowledge Dr. Xiaoji Niu for his guidance. Last, but not least, we would like to recognize the work done by Dr. Eun-Hwan Shin for the MMSS research group that was the starting point of all our work at MMSS.

The authors will be grateful to the readers who offer ideas for improvements or point out any errors or omissions in the book.

Zainab Syed
Calgary, Alberta, Canada
August 2010

1

Microelectromechanical Systems (MEMS)

1.1 Introduction

Microelectromechanical systems (MEMS) are the integration of mechanical elements, sensors, actuators, and electronics on a common substrate. MEMS bring together silicon-based microelectronics and micromachining technologies on a common platform [1, 2]. MEMS sensors gather information about the environment by measuring mechanical, thermal, biological, chemical, optical, and magnetic phenomena. The electronics then process the information derived from the sensors and through a decision-making capability direct the actuators to respond by moving, positioning, regulating, pumping, and filtering, thereby controlling the environment for some desired outcome or purpose [3]. MEMS are miniaturized, low-cost, low-power, silicon-based sensors. Since they are compact and not a classified product, these sensors are becoming popular in consumer-grade navigation systems. MEMS technology represents a progressive area of the electrical, mechanical, biomedical, aviation, and automotive industries. These devices allow the replacement of conventional bulky mechanical devices at a fraction of the cost, with enhanced durability, mass-production capability, and lower power consumption [1, 3]. Market analysis reports that the MEMS market grew by 9%, bringing in an impressive $7 billion in sales in 2007, and the automotive industry continues to drive the MEMS market [4].

MEMS sensors-based inertial navigation provides the positioning information when the most commonly used Global Positioning System (GPS) is lacking. Whether it is finding the attitude information to complete the navigation solution, or bridging short GPS outages, the MEMS sensors are proving their ultimate worth in the newly emerging seamless navigation field. Furthermore, MEMS gyroscopes (or gyros) have the potential to add value to consumer electronics devices by providing image stabilization, pedestrian navigation, and improved user interfaces. Moreover, MEMS inertial sensors are now being used in many other applications.

In May 2006, Nintendo and Analog Devices announced the incorporation of Analog Devices' ADXL330 *i*MEMS accelerometer into Nintendo's Wii console [5]. The accelerometer helped Nintendo raise video gaming to a new level. Similarly, Apple's iPhone uses an STMicroelectronics accelerometer to sense whether the iPhone is in a portrait or landscape position and the iPhone rotates the screen automatically.

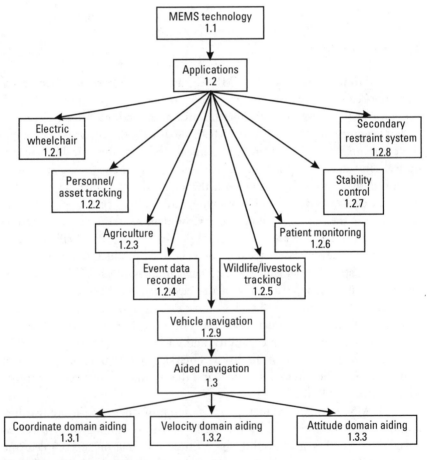

Figure 1.1 Flowchart of Chapter 1.

The applications for MEMS sensors do not stop with the examples provided above. With these sensors' three desirable features (i.e., low cost, low power, and small size), the applications are endless. An organization chart for Chapter 1 is provided in Figure 1.1 to demonstrate the basic structure of this chapter. This chapter briefly illustrates different application areas with emphasis on vehicle navigation applications. Different aiding sources for the inertial sensors navigation are also discussed to give users an overview of the organization of this book.

1.2 Different Applications of MEMS Devices

MEMS is an enabling technology with a massive global market volume. It is worth $12 billion in sales in 2004 and is expected to reach $25 billion in 2009 [6]. A small portion of this MEMS market will support inertial sensor technology. There are many different areas where inertial MEMS-grade sensors have been successfully applied to improve the quality of human life.

1.2.1 Electric Wheelchairs

Electric wheelchairs are important tools that aid in the mobility of disabled and elderly individuals [7, 8]. For elderly individuals, an intelligent wheelchair must be user independent and should assure collision-free travel, aid the performance of specific tasks (i.e., passing through doorways), and autonomously transport the user between locations. A smart wheelchair typically consists of a standard power wheelchair base to which a computer and a collection of sensors like inertial, vision, and so forth have been added, along with a seat placed over the base. The inertial sensors-based assistive wheelchair navigation system employs different operating modes and has the capability of automatically adapting the system based on human behavior.

1.2.2 Personnel Tracking and Navigation

Personnel tracking is an area where MEMS-based inertial sensors are becoming more and more popular. The inertial sensors provide important positioning information about personnel in GPS-denied environments [9, 10]. For example, tracking firefighters inside a building is not possible with GPS-only navigation. There are systems that are targeted towards the safety of the firefighters that contain accelerometers, gyroscopes, and other dead-reckoning sensors.

The use of such MEMS-based systems is also becoming an integral part in army personnel tracking systems. The most vital requirement for sensor-based navigation is the orientation of the sensors with respect to a known reference frame [11], which is achieved by embedding the sensors in either the shoes or helmets of the personnel. In inertial navigation systems for on-foot civilians, the orientation can be a limiting factor. This is the single most important obstacle towards the realization of a commercial sensor-based navigation system.

1.2.3 Agriculture

Precision agriculture is becoming another area where MEMS-based sensors can be quite helpful (e.g., the GreenStar2 system developed by John Deere). GPS provides absolute positions and is usually available in rural settings. In agricultural applications, roll and pitch estimation is important for the machinery platform. The integrated attitude and position information is the basis of the smart agricultural systems. The smart systems optimize the use of land by giving precise information to the user, and can be used in the application of fertilizers, herbicides/insecticides, harvesting, and land preparation for crops. In addition, integration of the navigation solution with land management systems also provides vital information for the treatment of land for maximum productivity.

MEMS-based tilt sensors, which are essentially accelerometers, can be used with the GPS receiver to obtain the required attitude and navigation information. The tilt of the machine is important, as the roll and pitch within farms can be quite significant. If the roll and pitch is not compensated, the application areas will either overlap or skip. In higher-accuracy agricultural products, three orthogonal accelerometers and three orthogonal gyroscopes are used. The system is easy to install for older farm machinery; some of the new machinery comes with this system. The user is notified about the exact application location by LEDs on the display module. In addition to the cost of the system, the user is also required to subscribe to the Differential Global Positioning System (DGPS) service for precise GPS positioning. However, all of the costs are offset by the reduction in the amount of work and chemicals used on farms.

1.2.4 Event Data Recorder

Over the past several years, there has been considerable interest in the community regarding possible safety benefits from the use of event data recorders (EDRs) in motor vehicles [12]. EDR mainly collects vehicle crash information, which includes information such as vehicle acceleration/deceleration data, driver inputs (e.g., braking and steering), and vehicle systems status. The EDR monitors several of the vehicle's systems, such as speed, brakes, and safety systems. The information obtained from EDRs aids investigations of the causes of crashes and injury mechanisms. This makes it possible to identify and address safety problems. The information collected can be used to improve motor vehicle safety. An EDR system includes angular rate sensors for monitoring rotational movement about three axes along with three linear accelerometers, or one triaxial linear accelerometer to track linear movement along the same three axes.

1.2.5 Wildlife and Livestock Tracking

Animal tracking is becoming common to study habitats and to identify movement behaviors [13, 14]. Usually animal tracking is done using GPS collars, but this kind of tracking is not sufficient as animals can be at places where a GPS signal may not be available. Most recently, animal tracking is done using the inertial sensors for continuous positioning and habitual information. In addition to the inertial dead-reckoning sensors, the tracking units also consist of digital cameras for visual aiding of the researchers. This particular system is used to track grizzly bears to understand their behavior and habitat.

Research is also ongoing in the field of livestock tracking to better understand grazing behavior in conjunction with various environmental conditions. Special collars of GPS and inertial sensors integrated with wireless sensor networks are deployed to monitor the movement of livestock. GPS and Internal Navigation System (INS) are used in their usual manner and the wireless sensor network allows communication among different nodes for the routing of important data.

1.2.6 Patient Monitoring

MEMS-based sensors are also used in some medical applications and patient monitoring [15]. These sensors are well-suited to monitor epilepsy as they can be used to quantify the intensity, frequency, and duration of movements. Research is underway to classify and correct motion disorders due to Parkinson's disease using MEMS inertial sensors.

In addition, MEMS accelerometers are commonly used to detect falling by high-risk or elderly patients who lose control. This kind of patient monitoring allows the patient more freedom and reduces the burden of the primary care provider. The patients can function more independently as any unusual event (from accelerometers signals analysis) is communicated to the emergency responder wirelessly.

1.2.7 Electronic Stability Control

Most recently, vehicles are coming equipped with an electronic stability control system [16]. The system consists of a MEMS gyroscope and accelerometers that can detect the skidding or loss of control by comparing the vehicle's actual motion with the steering angle. The system then applies brakes to individual wheels to bring the vehicle back into the control of the user. These systems can be intelligent enough to respond to emergency situations like a crash, skid, and so forth. Due to the effectiveness of the system, this technology is now becoming a fundamental component of all new vehicles.

1.2.8 Supplemental/Secondary Restraint System

Since 1998, a supplemental restraint system (SRS) is mandatory in all North American vehicles. The SRS consists of a deceleration (acceleration) sensor, which causes airbags to be deployed in the case of a frontal collision. The sensor module consists of MEMS accelerometers with a microprocessor. In general, if the force of collision exceeds +/– 7g, the SRS will deploy the airbags; for redundancy and robustness, other MEMS accelerometer modules are placed in the front panel of the vehicle, which are also used in the feedback of the deployment.

In recent years, SRS has become a more evolved and robust system. The system that was initially designed for frontal collision is now capable of providing protection to the vehicle occupants in case of rollover or side impact using side-impact or curtain airbags. These airbags, on deployment, stay inflated to provide maximum protection.

1.2.9 Land Vehicle Navigation

For several years, the integration of a high- or medium-quality INS with GPS has been implemented for accurate land vehicle-based surveying [17]. However, these grades of inertial systems are limited by their significant size and high cost ($25,000–150,000). With federal access regulations, the use of such systems is permitted only for authorized personnel. In the last decade, the demand for land vehicle navigation services has grown rapidly. To meet the navigation demand, many studies have been directed towards developing low-cost MEMS inertial systems. Advances in MEMS technology have made it possible to produce chip-based inertial sensors that are small, lightweight, consume very little power, and are reliable for land vehicle navigation and tracking systems.

However, with these advantages comes the disadvantage of low navigation accuracy. MEMS-based sensors provide accurate navigation data over short time intervals but suffer from accuracy degradation over extended time periods due to the combined effects of sensor errors like noises, biases, drifts, and so forth [18]. Due to the integration of inertial measurements (as given in Section 1.1), any residual errors will accumulate, resulting in position and orientation errors. Therefore, to limit these MEMS sensor errors, frequent updates from external aiding sources like the GPS, odometer, barometer, and so forth are required.

1.3 Aided MEMS-Based Inertial Navigation

Many different aiding sources are available to correct INS standalone system errors and are described below. The aiding sources can be classified as follows: aiding in coordinate domain, aiding in velocity domain, and aiding in attitude domain.

1.3.1 Aiding Sources in Coordinate Domain

These aiding sources mainly provide positioning information to compensate or reduce standalone MEMS sensor errors. Some of these sources are repeated for other domains.

1.3.1.1 Global Positioning System (GPS)

By far the most popular external aiding source for inertial data is GPS. GPS provides long-term accurate and absolute measurements and is well suited for integration with other sensors [19, 20]. However, this system does not work well in all environments as the GPS receiver requires a direct line of sight to four or more satellites. An integrated system combines the advantages of both systems and thus produces accurate and uninterrupted navigation results. In such an integrated system, GPS is typically used to provide accurate absolute positioning information to frequently update the INS; the INS is used to offer a short-term solution during GPS signal outage periods. DGPS is a technique used to locally improve the performance of GPS. A local base station receives the GPS signals and calculates offsets for each satellite with respect to the surveyed position. This set of offsets is then communicated to the local GPS receivers to correct their measurements for common ionospheric errors. The future deployment of the European satellite system, Galileo, will improve global coverage even further.

Other satellite-based augmentation systems (SBAS) based on DGPS principles have also been developed. Three main SBAS are the WAAS program sponsored by the Federal Aviation Administration (FAA) in the United States, EGNOS sponsored by the European Space Agency, and MSAS sponsored by the Japanese Civil Aviation Bureau.

1.3.1.2 Cell Phone Positioning

Cellular phone signals from two or more cell phone towers can be used to obtain the location of the handset. The positioning is absolute but it is not accurate as usually there is no direct line of sight to the cell phone towers of the handset. Signal triangulation is performed, which provides the approximate area of the user. In the absence of other more accurate updates in the position domain, cell phone signal positions can be used.

1.3.1.3 WiFi Positioning

WiFi technology is developing rapidly; it operates on unlicensed industrial, scientific, and medical bands, which lead to the opening of other bands. WiFi positioning systems may not require a direct line-of-sight path to the WiFi access points (AP) and are least affected by multipath, unlike GPS. By means of integrating a WiFi positioning system with the inertial data for an indoor scenario and with GPS and inertial data for outdoor environments, a real-time uninterrupted positioning may be performed. In WiFi, measurements are taken from all the hotspot signals in range and from prior information [21] to obtain the best solution. The main drawback of WiFi technology is the essential survey required to map the WiFi hotspots both at the initialization of the system or at every major change in the local environment.

1.3.1.4 Ultrawideband Positioning

Ultrawideband (UWB) signals provide accurate positioning capabilities because they use nanosecond pulses of radio energy, which are ideal for accurate timing information [22]. Consequently, they can be quite useful in providing updates for the inertial systems. The wide bandwidths (typically 500 MHz) offer the advantages of reliability, accuracy, and wall penetration capabilities. Furthermore, spreading the information over the wide spectrum keeps the power spectral density low and avoids interference with other systems. The downside of UWB is usage range, which is generally limited to approximately one kilometer with high-gain antennas, and for the fastest data rates, can be measured in tens of feet. UWB trades bandwidth for distance, so longer links are slower.

1.3.1.5 Radio Detection and Ranging (RADAR)

RADAR is a system that uses electromagnetic waves to identify the range, altitude, direction, or speed of both moving and fixed objects such as aircraft, ships, motor vehicles, weather formations, and terrain [21]. These systems use the reflected property of radio signals. The position of the target is estimated by measuring the angle of the signal reflection and the transit time of the transmitted signal to the target, and then back from the target to the receiver. RADAR can be useful in two different ways: first, the information provided by the system can be used to avoid any obstacles; second, the distance to a known location can be used to update the positioning information of an inertial system. The major limitation of this system is in separating unwanted signals in order to focus on the actual targets of interest.

1.3.1.6 Radio-Frequency Identification (RFID) System

Radio-frequency identification (RFID) is an automatic identification method, relying on storing and remotely retrieving data using devices called RFID tags or transponders. An RFID tag is an object that can be applied to or incorporated into a product or person for the purpose of identification and tracking using radio waves. RFID tags generally contain at least two parts: an integrated circuit (for storing, processing, and modulating/demodulating a radio-frequency (RF) information signal) and an antenna (for receiving and transmitting the signal). However, a primary RFID concern is the illicit tracking of RFID tags along with security and privacy concerns for the personnel.

1.3.1.7 Proximity Detection Techniques

A more simplistic and less accurate positioning technique is proximity detection that relies on the absence or presence of a signal when the receiver and transmitter are within a certain degree of proximity. The user is within the proximity range if the power level of the received signal falls within a predefined threshold value. It is a very simple and cheap approach and does not require any extra equipment for positioning. Nevertheless, the accuracy is poor because the environment can have different interfering signals of comparable or more power levels.

1.3.1.8 Map Matching

Map matching is an important tool for navigation. Determination of exact positioning information may not be very helpful for a civilian user unless it is communicated in terms of a reference map. Usually a map is considered as a known reference and the commercially available GPS navigation systems show the positions on a digital map. Map matching is becoming increasingly important especially for those situations where the GPS quality is not good and the map-matching algorithm has to decide in which segment of the map the user is probably located depending on the available positioning data. Using a similar concept, the inertial positioning can be improved by using the information from the digital map in the absence of other aiding sources.

1.3.1.9 Barometer

For indoor applications, MEMS-based barometers can be very useful as they provide the relative change in height. These sensors can provide accuracies up to 2m, which makes these sensors attractive in identifying the change in elevation during standalone inertial navigation scenarios. In addition, the height sensor can be used in other forms of navigation when a GPS signal is not available.

1.3.2 Aiding Sources in Velocity Domain

Similar to the aiding sources for positions, there are certain aiding sources that provide velocity information to compensate or reduce standalone MEMS sensor errors.

1.3.2.1 Zero Velocity Updates (ZUPTs)

Zero velocity updates (ZUPTs) occur when a land vehicle stops, and they represent no dynamic region of the vehicle trajectory. The velocity errors at this time are due to short-term random disturbances, which can be reduced by averaging the observations. When a vehicle stops, corrections can be made to all navigation parameters that can significantly reduce position drift errors.

1.3.2.2 Nonholonomic Constraints

Nonholonomic constraints (NHC) can be used to further improve navigation parameter accuracy. NHC use the fact that a land vehicle cannot move sideways or vertically up or down, and, hence, these two velocity components should be zero. The major problem arises when the terrain consists of significant pitch angles.

1.3.2.3 Odometer

In some cases, it may be possible to obtain information from the odometer of the vehicle. The odometer will generally provide the speed of the vehicle along a single axis, which can be used to improve the accuracy of the navigation solution. For correct updates, the wheel circumference of the vehicle is an important parameter, and, therefore, without reasonable calibration for the wheel, the system will produce erroneous results.

1.3.2.4 GPS

GPS is also capable of providing velocity information.

1.3.3 Aiding Sources in Attitude Domain

These aiding sources mainly provide attitude information (i.e., roll, pitch, and azimuth) to compensate or reduce the attitude errors during standalone MEMS navigation.

1.3.3.1 Map Matching

Map matching is an important tool for navigation as mentioned earlier. Similar to a position update, a map can be helpful in finding the direction of the user. For example, if the navigation solution is drifting due to the integration of sensor errors, heading information from the map can correct that drift. In this case, if the map is showing a straight road, and due to the drift the inertial system is indicating a wrong heading, the attitude can be easily updated using the map information.

1.3.3.2 MEMS Magnetometers

The accuracy of the estimated attitude defines the accuracy of the overall navigation solution if inertial systems are used. For high-end sensors, only gyroscopes are used for attitude determination and their errors are compensated for by the Kalman filter solution. However, the drift in the gyroscope is overwhelming for low-cost MEMS sensors and needs other forms of attitude determination for long-term reliability of the navigation solution. The MEMS magnetometer measures the angle of the sensor's sensitive axis relative to a local magnetic field, which can then be used to compensate for the attitude errors in the navigation solution. In the absence of soft iron anomalies, the sensor will output Earth's background magnetic field as its initial value. In the case of soft iron anomalies, Earth's magnetic field lines of flux are distorted towards the anomaly. If the sensitive axis of the magnetic sensor points to the right and the soft iron anomaly is traveling left to right, then the magnetometer will initially see a decreasing field as more flux lines bend towards the oncoming anomaly and swings in the negative direction. This is one of the limiting factors for magnetometers.

Now the reader is familiar with some common applications and aiding sources for MEMS sensors. The question is how do MEMS inertial sensors work? As the focus of this book is on navigation, another important question is what is the minimum number of sensors that are required to design a navigation system?

References

[1]   Gad-el-Hak, M., *The MEMS Handbook, First Edition*, Boca Raton, FL: CRC Press, 2001, p. 1368.

[2]   Lin, L., and Pisano, A. P., "Silicon Processed Microneedles," *Journal of MicroElectroMechanical Systems*, Vol. 8, 1999, pp. 78–84.

[3]   Maluf, N., and Williams, K., *An Introduction to Microelectromechanical Systems Engineering, Second Edition*, Norwood, MA: Artech House, 2004.

[4]   NEXUS Market Analysis for MEMS and Microsystems III, 2005–2009, http://www.memsinfo.jp/whitepaper/WP18_MEX.pdf, visited on August 8, 2008.

[5]   Venkatesh, P., and Sullivan, F., "MEMS in Automotive and Consumer Electronics," *Sensors Weekly*, 2007, http://www.sensorsmag.com/sensors/Automotive/MEMS-in-Automotive-and-Consumer-Electronics/ArticleStandard/Article/detail/473909.

[6]   Clark, R.L., "An Overview of Developments in Precision Farming Systems," *IEEE/ASME International Conference on Advanced Intelligent Mechatronics (AIM97)*, Waseda University, Tokyo, Japan, June 16–20, 1997.

[7]   Mascaro, S., and Asada, H.H., "Docking Control of Holonomic Omnidirectional Vehicles with Applications to a Hybrid Wheelchair/Bed System," *IEEE International Conference on Robotics and Automation*, Vol. 1, 1998, pp. 399–405.

[8]   Wang, G., and Kao, I., "Intelligent Soft Contact Surface Technology with MEMS in Robotic and Human Augmented Systems," *IEEE International Conference on Robotics and Automation*, Vol. 2, 2000, pp. 1048–1053.

[9]   Alwood, R., and Faulkner, T., " Personal Tracking in GPS-Denied Environments Using Low-Cost IMUs," presented at *The WPI Precision Personal Locator Project Workshop*, Worcester, MA, Aug 6–7, 2007.

[10]   Foxlin, E., "Pedestrian Tracking with Shoe-Mounted Inertial Sensors," *IEEE Computer Graphics and Applications*, Vol. 25, No. 6, 2005, pp. 38–46.

[11]   Yun, S.C., and Park, P.G., "MEMS Based Pedestrian Navigation System," *Journal of Navigation*, Vol. 59, Issue 1, 2006, pp. 135–153.

[12]   Kowalick, T.M., "Pros and Cons of Emerging Event Data Recorders (EDRs) in the Highway Mode of Transportation," *IEEE VTC*, 2001, pp. 3037–3041.

[13]   Huhtala, A., et al., "Evaluation of Instrumentation for Cow Positioning and Tracking Indoors," *Biosystems Engineering*, Vol. 96, No. 3, 2007, pp. 399–405.

[14]   Hunter, A.J.S., Stenhouse, G., and El-Sheimy, N., "A New Technology Solution for the Study of Bear Movement and Habitat Use: The Animal Pathfinder^{TM}," *18th International Conference on Bear Research and Management*, Monterrey, Mexico, 2007.

[15]   Wu, Y.H., et al., "The Development of M3S-Based GPS Navchair and Tele-Monitor System," *Proceedings of the 2005 IEEE Engineering in Medicine and Biology 27th Annual Conference*, Shanghai, China, September 1–4, 2005, pp. 4052–4055.

[16]   Laine, L., and Andreasson, J., "Control Allocation Based Electronic Stability Control System for a Conventional Road Vehicle," *Intelligent Transportation Systems Conference (ITSC 2007)*, 2007, pp. 514–521.

[17]   Syed, Z.F., et al., "Civilian Vehicle Navigation: Required Alignment of the Inertial Sensors for Acceptable Navigation Accuracies," *IEEE Transactions on Vehicular Technology*, Vol. 57, No. 6, 2008, pp. 3402–3412.

[18]   Aggarwal, P., et al., "A Standard Testing and Calibration Procedure for Low Cost MEMS Inertial Sensors and Units," *Journal of Navigation*, Vol. 61, No. 2, 2007, pp. 323–336.

[19]   Grewal, M.S., Weill, L.R., and Andrews, A.P., *Global Positioning Systems, Inertial Navigation, and Integration*, New York: John Wiley & Sons, 2007.

[20]   Misra, P., and Enge, P., *Global Positioning System: Signals, Measurements, and Performance*, Lincoln, MA: Ganga-Jamuna Press, 2006.

[21]   www.ekahau.com

[22]   www.ubisense.net

[23]   Mannings, R., *Ubiquitous Positioning*, Norwood, MA: Artech House, 2008.

2

MEMS Inertial Sensors

2.1 Introduction

MEMS inertial sensors have been embraced by the auto industry in their quest to improve performance, reduce cost, and enhance the reliability of the vehicles. Inertial sensors are the position, attitude, or motion sensors whose references are completely internal except possibly for initialization [1]. An inertial sensor that measures the linear acceleration of a body about its sensitive axis is called an accelerometer. Similarly, a sensor that measures the rate of change of angular velocity or absolute angle in an inertial frame about its sensitive axis is called a gyroscope. Microaccelerometers alone have the second largest sales volume and gyroscopes will soon be mass produced at similar volumes [2]. Both accelerometers and gyroscopes can be used for a number of applications as mentioned in Chapter 1.

MEMS devices are traditionally made by either bulk or surface micromachining techniques. Bulk micromachining uses the full thickness of the wafer and forms the proof mass along with the suspension system by wet etching the silicon wafer. These devices were used in earlier days and are more expensive than their counterparts, surface micromachined devices. In bulk micromachined devices, mechanical components cannot be monolithically integrated with the supporting electronics and therefore a separate IC chip is needed to house the additional electronics. The separate IC chips result in an increase in the manufacturing cost. However these devices have larger mechanical components and are more accurate than the surface micromachined devices. One such example of the bulk micromachined device is the high-end navigation-grade accelerometers, developed by a commercial company like Northrop Grumman's LN 101E for military applications [3].

The technology capable of housing the electronics and the mechanical structures on the same chip is surface micromachining. Surface micromachining is based on the deposition and etching of different structural layers on top of the substrate. This technology preserves the wafer but deposits or etches additional layers above the wafer surface to achieve the desired shape. The layers are generally plasma etched. These low-cost accelerometers, developed by surface micromachining, are typically comb drive based and equipped with sophisticated on-chip signal conditioning electronics. The sensing elements in the comb drive are movable comb fingers, which are an order of magnitude smaller than bulk micromachined devices. However less thickness and mass of the comb fingers impose severe limitations on the performance of these accelerometers. For example, Analog Devices Inc. has manufactured a number of very low-cost sensors based on this surface micromachining technology.

Bulk micromachining process technology is currently undergoing a revolution driven by the incorporation of deep reactive ion etching (DRIE) of silicon as a replacement for orientation-dependent (wet) etching [4]. However, DRIE substrate looks like surface micromachining. DRIE uses high-density plasma to alternatively etch silicon and deposit etch-resistant polymer on side walls. This gives DRIE the capability of providing precise 90° sidewall angles along with randomly shaped linear geometries.

Lithographie Galvanoformung Abformung (LIGA) is a popular process to obtain high aspect ratio microstructures from metals and plastics. LIGA combines IC lithography, electroplating, and molding to obtain depth. Patterns are created in a substrate and then electroplated to create 3D molds.

The main focus of this chapter is to provide the user with the answers regarding how MEMS inertial sensors work, what the different types of sensors are, and most importantly, from the commercial point of view, how many sensors are required to perform navigation for a land vehicle. A complete flowchart of Chapter 2 is provided in Figure 2.1.

2.2 Accelerometers

An accelerometer measures the specific force in an inertial reference frame, which can be used to estimate the acceleration of the moving body. Therefore, an accelerometer in the direction of motion of a vehicle may be the most important as it should contain all the information. For a 3D motion, however, one would like to record the accelerations in all three directions to accurately model such a motion. This is discussed in more detail in Chapter 3.

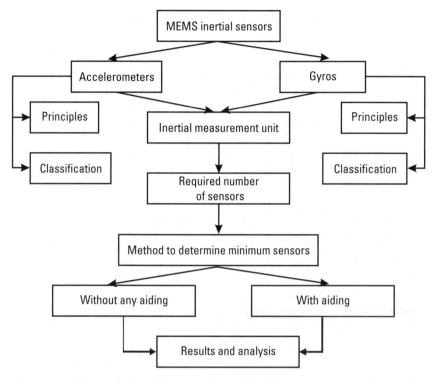

Figure 2.1　Flowchart for Chapter 2.

2.2.1　Working Principle for MEMS Accelerometers

An accelerometer operates by measuring the inertial force generated when a proof mass accelerates [5, 6] as illustrated by Figure 2.2. An accelerometer is made of at least three components, namely a proof mass, a suspension to hold the mass, and a pickoff, which relates an output signal to the induced acceleration.

The equilibrium position of the proof mass is calibrated for zero acceleration. Acceleration along the sensitive axis causes the proof mass to be displaced from its equilibrium position. The amount of resultant displacement from the equilibrium position is sensed by a pickoff, and then scaled to provide an indication of acceleration along this axis.

Acceleration in the positive direction will cause the proof mass to move downwards with respect to the equilibrium position. Imagine that the accelerometer is sitting on a bench in a gravitational field. It is displaced downward

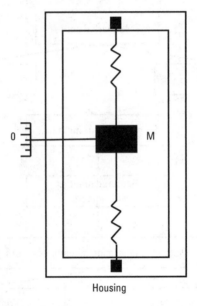

Housing

Figure 2.2   Basic accelerometer structure.

(with respect to the equilibrium position) due to its gravitational force as shown in Figure 2.3. The arrow in Figure 2.3 indicates the input-sensitive axis along with the positive direction of acceleration. As the accelerometer is stationary sitting on the bench, acceleration in the opposite direction must be exerted by it. Therefore, it exerts minus gravitational acceleration.

Accelerometers measure specific forces rather than these gravitational forces. The output of an accelerometer (acceleration minus gravitation) is called specific force and is given by (2.1) [6].

$$f = a - g \qquad (2.1)$$

where

f = specific force;

a = acceleration with respect to the inertial frame;

g = gravitational acceleration.

The units of these acceleration and force terms can be m/s^2 or some derivative of this unit such as mGal, which is 10^{-5} m/s^2 or μG that equals 0.981 mGal.

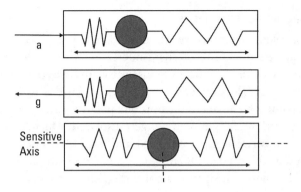

Figure 2.3 Displacement of proof mass.

For the above case, the accelerometer is sitting on a bench and is at rest (i.e., *a* is 0). The force on the accelerometer is actually the reaction force of the bench, which is equal to −*g*. Therefore, the accelerometer output for our example is *f* = −*g*. Another way of looking at it is to imagine the accelerometer dropping in a vacuum (free fall). In this case *f* = 0 because *a* = *g*. To navigate with respect to the inertial frame, it is required to determine the body acceleration *a*. Thus, in the navigation equations, we convert the output of the accelerometers from *f* to *a* by adding *g*.

2.2.2 Classifications of Accelerometers

A common way to classify accelerometers is based on the types of transduction used for converting the mechanical displacement of the proof mass to electrical signal. Some of the common principles are piezoresistive, capacitive sensing, piezoelectric, optical sensing, and tunneling current sensing.

2.2.2.1 Piezoresistive Accelerometers

These are the first micromachined and commercialized inertial sensors [2]. In this method, when the frame moves, suspension beams elongate or shorten, which changes the stress and resistivity of embedded piezoresistors. The change in resistance can be easily evaluated using standard bridge techniques. The main advantage is the simple fraction process involved for manufacturing these types of sensors. However, these sensors have a large undesired temperature sensitivity and small output sensitivity. Typical performance levels for these devices are a sensitivity of 1–3 mV/g, 5–50g dynamics range, and uncompensated temperature coefficient of 0.2%/°C.

2.2.2.2 Capacitive Accelerometers

Capacitive accelerometers incorporate an air-damped, opposed-plate capacitor as their sensing element. This mechanism creates a very stable, accurate measurement device, which is inherently insensitive to base strain and transverse acceleration effects. The differential capacitors are formed by using the proof mass as the common contact of the opposite polarity electrodes. On applying external acceleration [2, 7], proof mass gets displaced, which, in turn, changes the capacitance between the proof mass and fixed electrodes. These devices have a high sensitivity, good DC response, low temperature sensitivity, linear output, and low power dissipation along with their simple structure. However, these devices are prone to external electromagnetic interference. Moreover the sensitivity and resolution of capacitive accelerometers depend on the area of capacitive plates. With the decrease in device size, the sensitivity and resolution of the unit becomes limited.

2.2.2.3 Piezoelectric Accelerometers

Here the force or acceleration applied to the proof mass is converted to an electrical signal by the piezoelectric material. The obtained voltage is directly fed to the electronics with no conversion required. But these devices have a leakage issue, which deteriorates the DC response.

2.2.2.4 Tunneling Accelerometer

A constant current is maintained between a tunneling tip attached to the microstructure and its counter electrode. As the tip is brought closer to its counter electrode, a tunneling current is established and remains constant if no displacement occurs. When displacement occurs, a voltage is adjusted to maintain the current at a constant level. The voltage is a measure of applied acceleration. These devices have very high sensitivity in the range of 10^{-7}g, but suffer from significantly low-frequency noise.

Another criterion to classify accelerometers is based on the type of control loop used (i.e., whether the system is open- or closed-loop) [7]. In a closed-loop system, a force is fed back to the proof mass to counterbalance the inertial force. The measure of this feedback force gives a measure of the input acceleration. These sensors have more bandwidth, linearity, and dynamic range than their open-loop counterpart. However, the costs of these systems are very high due to the electronics' complexity. Consequently, most commercial accelerometers are open-loop systems.

2.3 Gyroscopes

To fully describe the motion of a body in 3D space, rotational motion as well as translational motion has to be measured. Sensors that measure angles or angular rates with respect to an inertial frame of reference are called gyroscopes [8].

2.3.1 Principle of MEMS Gyroscopes

The angular rate sensors or gyroscopes provide the change in angle with respect to an initially known orientation. Almost all reported micromachined gyroscopes use vibrating mechanical elements, called proof mass, to monitor rotational motion. They don't have any rotating part that requires bearings and hence can be miniaturized to reduce the overall cost of the sensor. These vibratory gyroscopes are based on the principle of energy transfer between two vibration modes of a structure caused by Coriolis force as shown in Figure 2.4. In this figure, x^e and y^e, represent the *x-axis* and the *y-axis* of the Earth's reference frame.

Coriolis force is named after a French scientist and engineer G.G. de Coriolis. Coriolis force is an apparent acceleration needed to hold Newton's

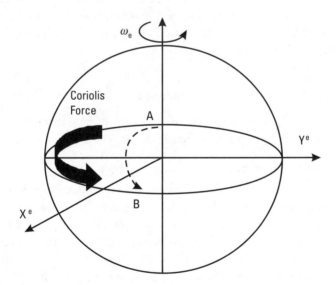

Figure 2.4   Coriolis force.

laws of motion in rotating reference frames. Sometimes this force is called a fictitious force (or *pseudo force*), because it does not appear when the motion is expressed in an inertial frame of reference. Due to the effect of the Coriolis force, a body moving from point A to B on the surface of the Earth will follow a curved line instead of a straight line as demonstrated in Figure 2.4. At a given rate of rotation of the observer, the magnitude of the Coriolis acceleration is related to the body velocity, the angle between the direction of motion of the body, and the axis of rotation as given by (2.2).

$$\vec{a}_c = -2\vec{w} \times \vec{v} \qquad (2.2)$$

where $\vec{v}$ is the body velocity in the rotating frame and $\vec{w}$ is the angular velocity of the rotating system. Equation (2.2) implies that the Coriolis acceleration is perpendicular to both the direction of the velocity of the moving mass and the rotation axis. All vibrating gyroscopes employ the Coriolis effect for measuring the angular rates. In comparison to accelerometers, gyroscopes are a challenging technology and are in the development stages. The main difficulty is that the sensing element must be able to move or vibrate in two degrees of freedom, one for the excited or driven mode (in which the body moves with velocity $\vec{v}$), and other for the sense mode (which experiences Coriolis force). Thus, gyroscopes can be described as a resonator along the drive direction and as an accelerometer along the sensing direction. Furthermore, the Coriolis coupling between the drive and sense mode is very weak, thus resulting in the use of mechanical amplifications.

2.3.2 Classification of MEMS Gyroscopes

For gyroscopes, there are three common designs of fabrication. Each will be discussed in the following subsections in detail [7, 8].

2.3.2.1 Tuning Fork Gyroscopes

Tuning fork gyroscopes are a common example of vibratory gyroscopes. The tuning fork consists of two tines, which are connected together at the junction as shown in Figure 2.5. The tines are resonated differentially along two axes by a suspension system. When this system is rotated, Coriolis force is generated, which causes a differential sinusoidal force to develop on the individual tines orthogonal to the main vibrations. The actuation mechanism used for producing resonance on the tines can be electrostatic, electromagnetic, or piezoelectric. Also to sense the Coriolis-induced vibrations in the sense mode, capacitive, piezoresistive, or piezoelectric methods can be used.

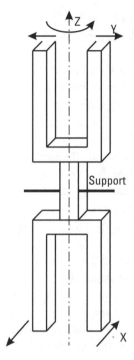

Figure 2.5   Tuning vibratory gyroscope.

2.3.2.2 Vibrating Wheel Gyroscopes

As the name indicates, these gyroscopes have a wheel that vibrates about its axis of symmetry. The wheel rotation about the symmetry axis results in the wheel tilting, which will produce a change in the measurement. This change can be detected in a number of ways, such as implementing electrostatic, electromagnetic, piezoelectric, piezoresistive, or capacitive circuits.

2.3.2.3 Wine Glass Resonator Gyroscopes

Wine glass gyroscopes, or hemispherical gyroscopes, are fabricated from fused silica rather than silicon. In the wine glass gyroscope, the resonance of a resonant ring is the measured variable, and the positions of the nodal points indicate the rotation angle.

Now that we have seen the major classifications of MEMS-based inertial sensors along with the working principle of these sensors, we are ready to take this topic to the next level. We will now discuss the number of inertial sensors that are required for land vehicle navigation.

2.4 MEMS Inertial Sensors for the Most Economical Land Navigation

An inertial sensors-based navigation system generally consists of three gyroscopes and three accelerometers. Gyroscopes sense angular velocity ω_{ib}^b, which is the rotation of the b-frame with respect to the i-frame, measured in the b-frame. Accelerometers measure the specific force f^b in the b-frame. This configuration is required for air or marine navigation systems where attitude angles (i.e., roll, pitch, and azimuth changes) are significant. However, ground vehicle navigation and positioning are constrained mainly to two dimensions. This chapter presents and compares different sensor configurations that can be used for vehicle navigation, taking into consideration the robustness of the navigation or positioning results, and hardware cost of the inertial measurement unit (IMU).

Heading rate information and a vehicle's longitudinal acceleration are of great significance for land navigation and must be included in any sensor configuration. If a heading gyroscope and a forward/backward sensing accelerometer are present in a land vehicle inertial system, theoretically, the system should be able to provide the navigation information.

One method to reduce the cost of the inertial system is to reduce the number of sensors to the minimum sensors required for robust navigation. In this chapter, five possible sensor combinations are provided as illustrated by Figure 2.6. Initially, a full inertial sensor system consisting of a triaxial accelerometer and three orthogonal gyroscopes (full IMU) is studied. The full IMU is capable of providing three-dimensional motion as it covers all three dimensions and consists of six sensors, which makes it the most expensive option. Today, dual-axis MEMS gyroscopes [9] are coming onto the market that can sense the motion in two orthogonal dimensions. There are times when the pitch angles for the vehicle can be quite significant in land navigation, such as a driving route passing through mountainous terrain. The addition of a gyroscope to sense those pitch angles can prove to be beneficial, and, hence, the second configuration considered consists of a dual-axis gyroscope and triaxial accelerometer (DG-TA). The next sensor configuration is even more economical as it uses only two sensors [i.e., a single axis gyroscope to measure the heading rate and a triaxial accelerometer (SG-TA)]. Since dual-axis accelerometers are also available in the market, another economical sensor format can utilize such dual axis accelerometers. In this configuration (SG-DA), the dual axis accelerometer is used to sense the specific forces in the horizontal plane as the vertical direction

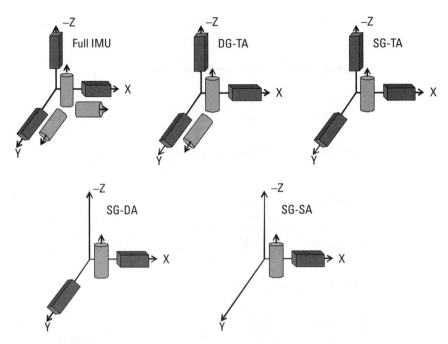

Figure 2.6   The five sensor configurations.

senses mostly the gravity. The last case is the one that consists of a single-axis gyroscope for heading rate measurement and a single axis accelerometer in the forward direction (SG-SA). It is the simplest configuration that covers two important directions for the land vehicle navigation.

The typical inertial dataset, as shown in Figure 2.7, clearly shows that for navigation on a level road, the roll, pitch angles, and vertical velocities are negligible. Mostly, a vehicle moves forward on a road, which is evident from the signals of the forward accelerometer. The lateral accelerometer is also providing trajectory information as it measures the centripetal accelerations that are significant during vehicle turning. A vertical gyroscope is necessary to measure the heading of the vehicle as mentioned earlier. Hence, for a well-paved and level road, the roll and pitch gyroscopes (i.e., gyroscopes in forward and lateral directions) and vertical accelerometer will provide minimal information.

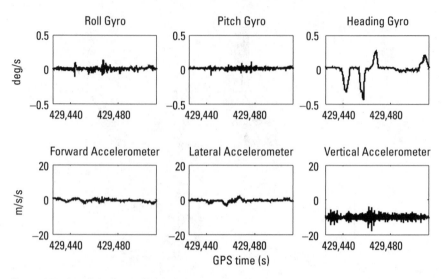

Figure 2.7   Inertial sensor signals for a typical land vehicle trajectory.

Despite using a minimum number of sensors, vehicle navigation can still implement velocity aiding and NHC to improve the navigation solution [10].

2.5 Method to Compute Minimum Sensors

There are three possible ways to deduce the minimum number of sensors for land navigation. The first method is to collect data with the required number of sensors and then process the data with simplified navigation equations (Figure 2.8). In Figure 2.8, f_x, f_y, and f_z represent the output of accelerometers in the x, y, and z directions, while w_x, w_y, and w_z represent gyroscope output in three directions.

This method is most realistic, as it requires both hardware and software modifications. In addition, extensive data collection is required to obtain the results for different hardware modules. Generally, since hardware is more costly and difficult to modify than software, it would be more favorable to only work with the software modifications. This brings in the second method that requires the simplification of the navigation equations. As before, the hardware remains the same (i.e., consists of all six sensors). In this case, only data from the required sensors are provided to the simplified equations as if the other sensors are not part of the system. This method also needs extensive software development.

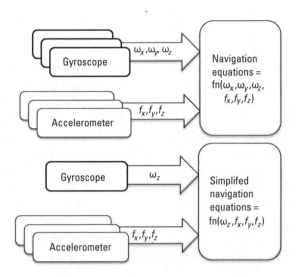

Figure 2.8 Software modification example for full IMU and SG-TA sensors.

The third method requires the least amount of work as it can be implemented at the integration level. Usually a Kalman filter (KF) is used to optimally integrate the inertial sensors data with the available aiding sources. This method modifies the signals and their stochastic properties at the integration level so that the KF does not use the unwanted signal for integration. Due to its simplicity, this method was used to deduce the minimum number of sensors. To verify our methodology, two different datasets collected in Calgary, Alberta are used for the analysis.

The original version of the integration KF has processing capabilities for the full IMU data consisting of six sensors outputs (i.e., three accelerometers' and three gyroscopes' signal output). In the modified version, the noise and random walk [signal standard deviations, velocity random walk (VRW), and angle random walk (ARW)] of the omitted sensors are increased, the corresponding signals are assigned to pseudosignals and their weights (inverse of the variance covariance matrix) are set to zero.

The concept of pseudosignals is straightforward but there are certain requirements for using the pseudosignal. The new signal should satisfy the Gaussian white noise requirements for the KF. The systematic method to compute the pseudosignal Gaussian noise is by taking the spectral density, q, of the original signal. The q values are used to design the Q_k matrix, the process noise matrix, of the KF. The spectral density is a function of the standard deviation (STD) of the signal and the BW of the IMU [11].

$$\sqrt{q} = STD_{sensor} / \sqrt{BW_{IMU}} \qquad (2.3)$$

The corresponding noise values for the accelerometers and gyroscopes that need to be removed can be computed for the Q_k matrix. This matrix predicts measurements depending on the quality of the signals and thus it is important to satisfy the Gaussian white noise requirements. The accelerometers and gyroscopes used in this chapter are from the custom inertial unit (CIU) with a bandwidth of 40 Hz, which has signal standard deviations of 0.084 m/s^2 and 0.0035 deg/s (static portion of Figure 2.7). CIU is a MEMS IMU sensor triad that was designed by the Mobile Multi-Sensor Systems research group (MMSS) at the University of Calgary. Using these two values the spectral densities can be estimated as follows:

$$\sqrt{q_{accel}} = 0.084 / \sqrt{40} = 0.013 \text{ m/s}/\sqrt{s} \qquad (2.4)$$

and

$$\sqrt{q_{gyro}} = 0.0035 / \sqrt{40} = 5.5e^{-4} \text{ deg} / \sqrt{s} \qquad (2.5)$$

The KF may diverge if the values computed above are used when the actual errors of the pseudosignals are not white noise. To avoid such a problem, the noise spectral density is set to a much higher value to ensure that the filter does not provide any weight to the omitted sensors.

For the DG-TA configuration, the roll gyroscope is removed from the full IMU. As the data is from a full IMU, a pseudosignal(s) of 0.0 (in Figure 2.9) is assigned to the roll gyroscope and the corresponding weight (variance^{-1}) value is also set to 0. In addition to that, the ARW value for the roll gyroscope is increased by a factor (F), which is greater than 25 ($F > 25$). This is to ensure that the navigation KF does not use the measurement from this sensor and is equivalent to physically removing the sensor from the full IMU at the data processing level.

SG-TA configuration only has a heading gyroscope, and, similar to the first case, pseudosignals of 0.0 ($s = 0$) magnitudes are assigned to both roll and pitch gyroscopes and the signal weights are set to 0.0. The corresponding ARW values are increased for the two sensors ($F > 25$). The next set of simulations is for SG-DA sensor configuration. In this case, the vertical accelerometer signal is also removed by assigning it a pseudosignal value of −9.81 m/s/s ($s = −9.81$ m/s/s), which is in addition to the removal of the roll and pitch gyroscopes signals as mentioned in the SG-TA case. The corresponding weights of the

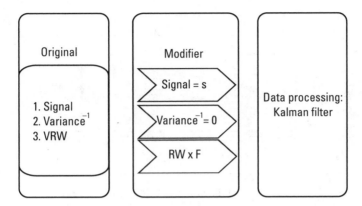

Figure 2.9 Partial sensor processing flowchart.

omitted sensors are set to 0s and the RW values are increased to ensure that the sensors are not used in the navigation KF.

The final configuration contains just one gyroscope, the heading gyroscope, and one accelerometer. In this case, pseudosignals of magnitude 0.0 replaced the original signals for the roll gyroscope, pitch gyroscope, and lateral accelerometer. The vertical accelerometer is assigned the pseudosignal of −9.81 m/s/s. The weights of the omitted sensors are reduced to zero and the corresponding RW values are increased.

2.6 Results and Analysis

This section is divided into two subsections. Section 2.6.1 presents the position drift errors with respect to the reference solution without the use of any extra aiding sources (e.g., NHC). Section 2.6.2 presents the same results but after NHC is applied during the test data processing.

2.6.1 Drift Errors Without NHC

The position drift errors without NHC for two datasets are given in Figure 2.10. The figure clearly demonstrates that the mean of the drift errors will increase as the sensors are removed from the inertial unit, despite the fact that land vehicle motion is constrained. Regardless of the different magnitude for the mean drift errors, the trend for the two datasets is similar.

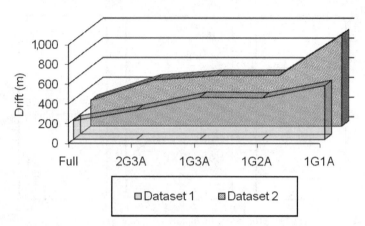

Figure 2.10 Average drift errors for the different sensor configurations for dataset 1 and dataset 2.

Figure 2.11 shows an example of the estimated roll and pitch with their standard deviations for full (top panel) and SG-DA (bottom panel) sensor configurations. The KF estimates the errors in roll and pitch but the error is much more random with a higher amplitude than the full sensor case. This is due to the higher RW values that are used in the simulations.

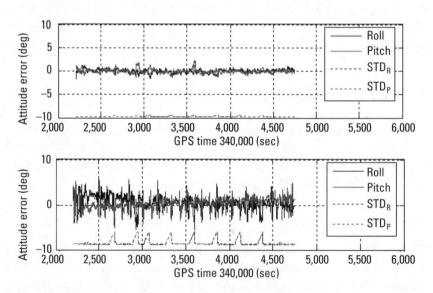

Figure 2.11 Estimated roll and pitch errors for full (top panel) and SG-DA (bottom panel) sensor configurations.

2.6.2 Drift Errors with NHC

The NHC is applied to both datasets and the mean drift errors are given in Figure 2.12. It is evident that the application of the NHC reduced the overall drift errors, and, therefore, application of these constraints is desirable for an accurate navigation system. NHC makes assumptions about the vehicle's velocity, and, therefore, a good alignment between the body frame of the IMU (*b*-frame) and the vehicle frame (*v*-frame) is required.

The dataset 1 has roll and pitch misalignments between the *b*-frame and *v*-frame, while dataset 2 has virtually no misalignment between the two frames as shown in Figure 2.13. Theoretically, the NHC for dataset 1 should not decrease the drift errors as effectively as for dataset 2 due to the underlying misalignment angles.

The NHC decreases the drift errors by only 37.62% for dataset 1 (dataset with small misalignments) and 81.13% for dataset 2 (with no misalignment). Full IMU outperformed any other partial sensor configuration, but it is interesting to note that the SG-DA sensor configuration produced better overall results as compared to the other partial sensor configurations for well-leveled roads used in the analysis. For dataset 1 the mean drift error for SG-DA is close to the full IMU; however, the DG-TA results are exhibiting slight degradation. For dataset 2, the mean drift errors for DG-TA, SG-TA, and SG-DA are similar to each other [10].

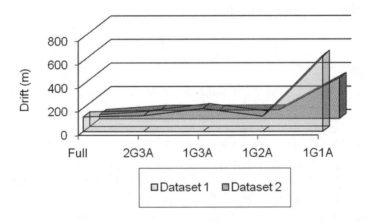

Figure 2.12 Average drift errors for the different sensor configurations for dataset 1 and dataset 2 using NHC.

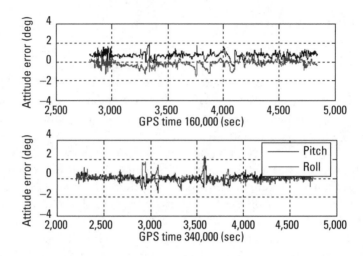

Figure 2.13   Roll and pitch misalignments for dataset 1 (top panel) and dataset 2 (bottom panel).

So far, we have discussed the minimum number of sensors along with a possible way to improve navigation using the MEMS-based inertial sensors. In addition, the principle and classification of the MEMS sensors was described. In the last section, we touched on some sensor errors of the MEMS inertial sensors that are required for navigation KF design. Are these the only kinds of errors for MEMS inertial sensors? What are the methods to actually reduce these errors? What are the ways to quantify all of the sensor errors for best navigation accuracies? Chapter 3 will answer these questions.

References

[1]   IEEE Std 1139-1988, "IEEE Standard Definitions of Physical Quantities for Fundamental Frequency and Time Metrology."

[2]   Barbour, N., et al., "Micromachined Inertial Sensors for Vehicles," *IEEE Conference on Intelligent Transportation System*, 1997, pp. 1058–1063.

[3]   Northrop Grumman, "Lufthansa Selects Northrop Grumman to Provide Next-Generation Air Data Inertial Reference Units," http://www.irconnect.com/noc/press/pages/news_releases.html, visited January 2009.

[4]   Gad-el-Hak, M., *The MEMS Handbook, First Edition*, CRC Press, 2001.

[5]   IEEE Std 1293-1998, "IEEE Standard Specification Format Guide and Test Procedure for Linear, Single-Axis, Non-Gyroscopic Accelerometers."

[6]   El-Sheimy, N., "Inertial Techniques and INS/DGPS Integration," ENGO 623-Course Notes, Department of Geomatics Engineering, University of Calgary, Canada, 2006.

[7]   Kraft, M., "Micromachined Inertial Sensors: The State of the Art and a Look into the Future," *IMC Measurement and Control*, Vol. 33, No. 6, 2000, pp. 164–168.

[8]   IEEE Std 952-1997, "IEEE Standard Specification Format Guide and Test Procedure for Single-Axis Interferometric Fiber Optic Gyros."

[9]   InvenSense Inc., http://www.invensense.com/news/060516.html, visited September 6, 2008.

[10]   Syed, Z., et al., "Economical and Robust Inertial Sensor Configuration for a Portable Navigation System," *ION GNSS 2007*, Fort Worth, Texas, Sept. 24–25, 2007.

[11]   El-Sheimy, N., "The Potential of Partial IMUs for Land Vehicle Navigation," *Inside GNSS*, 2008, pp.16–25.

3

MEMS Inertial Sensors Errors

3.1 Introduction

MEMS sensors are cost-effective units with small dimensions, but suffer from larger errors when compared with their mechanical predecessors [1, 2]. Low-cost MEMS inertial sensors exhibit high biases, scale factor variations, axis nonorthogonalities, drifts, and noise characteristics [3]. Gyroscope biases for MEMS IMUs lie in the range of 100°/hr and above while these biases are negligible for higher grade sensors (Figure 3.1). These errors build up over time, corrupting the precision of the measurements. Therefore, it is important to properly remove these errors before the start of navigation.

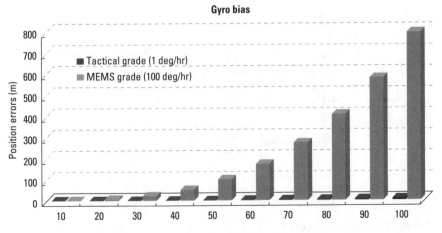

Figure 3.1 Position errors for tactical grade IMU with 1°/hr gyro bias as compared to MEMS-grade IMU with 100°/hr.

This chapter begins with a description of different sensor errors and their important classifications. Next, special methods to reduce the MEMS inertial sensor errors are provided. A complete measurement model for the inertial sensor is discussed that would account for all the possible error sources. A flow chart for Chapter 3 is provided in Figure 3.2.

3.2 Systematic Errors

Systematic errors are due to manufacturing defects and can be calibrated out from the data. Some systematic errors are given below.

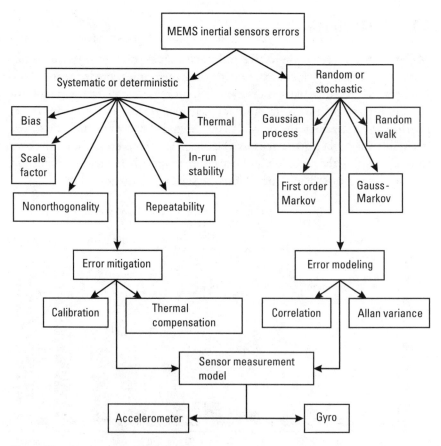

Figure 3.2   Flow chart for Chapter 3.

3.2.1 Bias

Ideally, when no input is applied to the sensor, the output signal received from the sensor should read 0 [2–4]. However, this is not the case, and an offset called the bias exists, as shown in Figure 3.3. The average of accelerometer or gyroscope output over a predetermined time that has no relation to input acceleration or rotation is called bias. The units for accelerometer bias are g or m/s^2 while that for gyroscope bias is °/h or °/sec. The low-cost MEMS unit biases for accelerometers are depicted in Figure 3.3. We can observe in Figure 3.3 that uncompensated biases are as huge as 19 m/s^2.

An uncompensated accelerometer bias introduces an error proportional to time (t) in the velocity and proportional to t^2 in the position as illustrated in (3.1) [4].

$$v = \int b_f \, dt = b_f t \Leftrightarrow p = \int v \, dt = \iint b_f t \, dt = \frac{1}{2} b_f t^2 \qquad (3.1)$$

where b_f is accelerometer bias, p is the position, and v is the velocity.

For example, consider a case where accelerometer bias is 8 m/s^2. If this bias offset is not removed from the measurements, it can generate a 400-m error in position after 10 seconds and 10-km error after 50 seconds using (3.1).

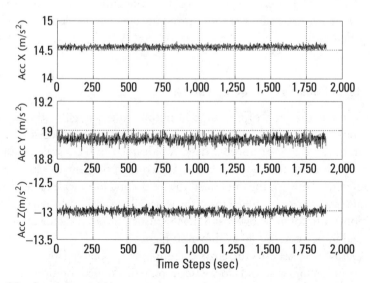

Figure 3.3 Accelerometer biases.

On the other hand, an uncompensated gyroscope bias in the x or y gyroscope will introduce an angle (in roll or pitch) error $\delta\theta$, proportional to time t as given in (3.2).

$$\delta\theta = \int b_w \, \mathrm{d}t = b_w t \tag{3.2}$$

where b_w is gyroscope bias.

The small tilt angle error will cause a misalignment of the INS, and therefore a projection of the acceleration vector in the wrong direction. This, in turn, will introduce proportional acceleration to $\delta\theta$ ($a = g\sin(\delta\theta) \approx g\,\delta\theta = gb_w t$) in one of the horizontal axes. Position and velocity errors, as a result of this angle, are calculated by (3.2).

$$v = \int a \, \mathrm{d}t = \int g b_w t \, \mathrm{d}t = \frac{1}{2} b_w g t^2 \tag{3.3}$$

$$p = \int v \, \mathrm{d}t = \iint \frac{1}{2} b_w g t^2 \, \mathrm{d}t = \frac{1}{6} b_w g t^3 \tag{3.4}$$

Therefore, a small gyroscope bias introduces quadratic errors in velocity and cubic error in the position as given by (3.3) and (3.4). If 8°/sec of gyroscope bias exists, it will introduce a 228-m error in the position after 10 sec and a 28.521-km error after 50 sec. Consequently, it is obvious that the outcome of gyroscope bias is more pronounced than accelerometer bias as time progresses.

3.2.2 Input Sensitivity or Scale Factor

Scale factor is the ratio of a change in output to a change in the intended input to be measured [4]. It is evaluated as the slope of the straight trend line that can be fitted to input-output data. An ideal sensor has a scale factor of 1; hence, any scale factor above or below 1 is contaminated with sensor errors. The difference between the imperfect scale factor and ideal scale factor is called the scale factor error. Scale factor error is usually expressed in parts per million (ppm). But for very low-cost MEMS sensors, the scale factor error can be expressed in percentage (%) as shown in Figure 3.4.

Gyroscopes scale factor error (Figure 3.4) is obtained by rotating the sensor counterclockwise and clockwise at some predefined rotation rates using the turntable according to angular rate test methodology (Section 3.3.2). Scale factor error of gyroscopes and accelerometers cause the same effect as

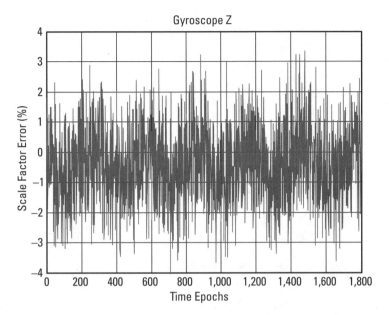

Figure 3.4 Scale factor error of a typical MEMS gyroscope.

that of the respective sensor bias. The gyroscope scale factor error causes position error proportional to t^3 while accelerometer scale factor error causes position error proportional to t^2.

3.2.3 Nonorthogonality/Misalignment Errors

Axes nonorthogonality is the error resulting from the imperfection in mounting the sensors at the time of manufacturing as illustrated in Figure 3.5 [5–7]. In addition, there could be a mounting misalignment error between the sensitive axes of the inertial sensors and the orthogonal axis of the body (Figure 3.6). In both cases, each axis is affected by the measurements of the other two axes.

The nonorthogonality errors are unitless, and generally expressed as parts per million. Consider an orthogonal triad of stationary accelerometers placed on a plane, which is tilted by a small angle $\delta\theta$ with respect to the y-axis (generally, called a roll) as shown in Figure 3.6. Due to this tilt, each accelerometer will measure a component of Earth's gravity g. In the z-direction, it will measure $f_z = g \cos \delta\theta$ and in the x-direction it will measure $f_x = g \sin \delta\theta$. Similarly, if the unit is tilted by a small angle $\delta\theta$ with respect to the x-axis (generally called a pitch), the z-axis experiences $f_z = g \cos \delta\theta$ and the y-direction measures $f_y = g \sin \delta\theta$.

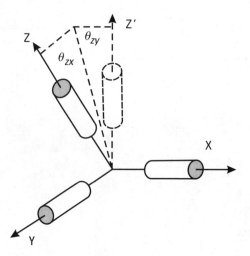

Figure 3.5   Nonorthogonality error due to manufacturing imperfections.

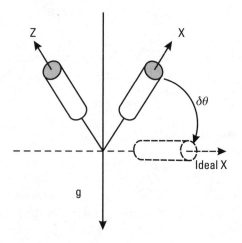

Figure 3.6   Mounting misalignment.

For very small angles, we can approximate $f_z = g \cos \delta\theta \approx g$ and $f_x = g \sin \delta\theta \approx g \delta\theta$. Thus a small misalignment error with respect to the y- and x-axis causes errors in the x- or y-directions, respectively, as given by (3.5).

$$v = \int g\delta\theta\,\mathrm{d}t = g\delta\theta t \Leftrightarrow p = \int v\,\mathrm{d}t = \iint g\delta\theta\,\mathrm{d}t = \frac{1}{2}g\delta\theta t^2 \qquad (3.5)$$

3.2.4 Run-to-Run (Repeatability) Bias/Scale Factor

Run-to-run bias or scale factor is the variation in bias or scale factor for each run. Basically every time the sensor is switched on, a slightly different bias or scale factor value is observed [1]. However, this error remains constant during a particular run. Note that these errors are significant only for low-cost sensors. The gyroscope bias and scale factor run-to-run errors along with the residual systematic bias or scale factor of a single-axis low-cost MEMS sensor are shown in Figures 3.7 and 3.8.

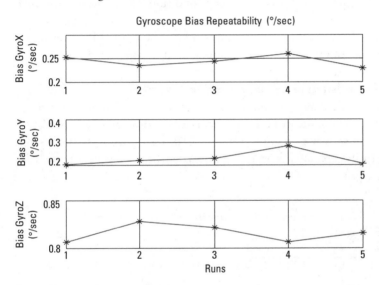

Figure 3.7 Gyro bias repeatability.

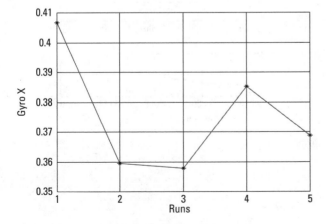

Figure 3.8 Gyro scale factor repeatability.

In these results, biases and scale factor errors are calculated according to the 6-position static test and angular rate test (Section 3.3) for each of the five runs separated by half-hour time intervals.

3.2.5 In Run (Stability) Bias/Scale Factor

This error occurs due to change in bias or scale factor during a run. Scale factor stability is the capability of the inertial sensor to remain invariant when continuously exposed to a fixed operating condition [5–7]. The deterministic portion of the in-run bias/scale factor error is caused by changing environmental conditions, such as temperature, and can be modeled. Nevertheless, the remaining component is hard to model and is considered a random or stochastic process.

Sensors are left running continuously for over two hours in the lab. Every hour the collected data is calibrated to obtain biases and scale factor drifts. The obtained biases and scale factor errors variations for low-cost MEMS gyroscopes are shown in Figures 3.9 and 3.10, respectively. As these errors cannot be easily separated from the sensor data, these are generally modeled as stochastic processes.

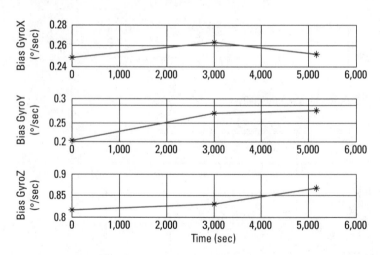

Figure 3.9   Gyro bias stability.

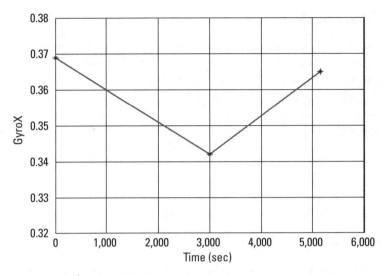

Figure 3.10 Gyro scale factor error stability.

3.2.6 Temperature-Dependent Bias/Scale Factor

This error is the variation of bias and scale factor errors with the change in sensor temperature [8, 9]. The temperature-dependent variations can be quite pronounced in very low-cost MEMS sensors and for certain applications may require reliable mitigation techniques. These variations are estimated by thermal testing procedures, which measure the discrepancy of basic sensor parameters when operated under different temperatures [1, 8].

3.3 Calibration of Systematic Sensor Errors

Calibration is the process of comparing instrument outputs with known reference information and determining the coefficients that force the output to agree with the reference information over a range of output values. Generally, calibration methods such as the 6-position static method and rate tests are used for this purpose. These methods used for calibrating IMUs are primarily designed for high-quality sensors, such as navigation or tactical grade IMUs. For low-cost automotive grade sensors, the above methods are modified using specialized instruments like turntables.

3.3.1 6-Position Static Test

The 6-position static and rate tests are among the most commonly used [8] calibration methods. The 6-position method involves mounting the inertial system on a leveled surface with sensitive x-, y-, and z-axes of the IMU pointing alternately up and down. For a triad of orthogonal sensors, this results in a total of 6 positions. The bias and scale factor errors are calculated using (3.6) and (3.7).

$$b = \frac{l_f^{up} + l_f^{down}}{2} \tag{3.6}$$

$$S = \frac{l_f^{up} - l_f^{down} - 2 \times K}{2 \times K} \tag{3.7}$$

where l_f^{up} is the sensor measurement when the sensitive axis is pointed upward, l_f^{down} is the measurement when the sensitive axis is pointed downwards and K is the known reference signal. For accelerometers, K is the local gravity constant, and for gyroscopes K is the magnitude of the Earth's rotation rate projection to the vertical at a given latitude. The current 6-position static tests cannot calibrate automotive-grade sensors as bias instability and noise levels completely mask the Earth's rotation rate projected reference signal. Furthermore, this standard calibration method cannot estimate the axes misalignments (nonorthogonalities) of the IMU.

To estimate the nonorthogonalities, an improved 6-position test can be performed, which takes into account all three types of errors. In matrix form, the output of a triad of sensors (e.g., accelerometers) can be represented as:

$$\begin{bmatrix} l_{ax} \\ l_{ay} \\ l_{az} \end{bmatrix} = \begin{bmatrix} m_{xx} & m_{xy} & m_{xz} \\ m_{yx} & m_{yy} & m_{yz} \\ m_{zx} & m_{zy} & m_{zz} \end{bmatrix} \begin{bmatrix} a_x \\ a_y \\ a_z \end{bmatrix} + \begin{bmatrix} b_{ax} \\ b_{ay} \\ b_{az} \end{bmatrix} \text{ or } \begin{bmatrix} l_{ax} \\ l_{ay} \\ l_{az} \end{bmatrix} =$$

$$\underbrace{\begin{bmatrix} m_{xx} & m_{xy} & m_{xz} & b_{ax} \\ m_{yx} & m_{yy} & m_{yz} & b_{ay} \\ m_{zx} & m_{zy} & m_{zz} & b_{az} \end{bmatrix}}_{M} \underbrace{\begin{bmatrix} a_x \\ a_y \\ a_z \\ 1 \end{bmatrix}}_{a} \tag{3.8}$$

Here the diagonal m elements represent the scale factors, the off diagonal elements are the nonorthogonalities, and the b components are the biases. By aligning the IMU using the standard 6-position method, the ideal accelerations would be measured as:

$$a_1' = \begin{bmatrix} g \\ 0 \\ 0 \end{bmatrix}, a_2' = \begin{bmatrix} -g \\ 0 \\ 0 \end{bmatrix}, a_3' = \begin{bmatrix} 0 \\ g \\ 0 \end{bmatrix}, a_4' = \begin{bmatrix} 0 \\ -g \\ 0 \end{bmatrix}, a_5' = \begin{bmatrix} 0 \\ 0 \\ g \end{bmatrix}, a_6' = \begin{bmatrix} 0 \\ 0 \\ -g \end{bmatrix} \quad (3.9)$$

Consequently, the design matrix (**A**) for the least squares adjustment will have the form:

$$\mathbf{A} = \begin{pmatrix} a_1' & a_2' & a_3' & a_4' & a_5' & a_6' \\ 1 & 1 & 1 & 1 & 1 & 1 \end{pmatrix} \quad (3.10)$$

The raw output of the sensors (in volts) constitutes the matrix **U**:

$$\mathbf{U} = \begin{bmatrix} u_1 & u_2 & u_3 & u_4 & u_5 & u_6 \end{bmatrix} \quad (3.11)$$

$$\text{where } u_1 = \begin{bmatrix} l_{ax} \\ l_{ay} \\ l_{az} \end{bmatrix}_{XaxisUP}, u_a = \begin{bmatrix} l_{ax} \\ l_{ay} \\ l_{az} \end{bmatrix}_{XaxisDOWN} \quad \text{and so on.}$$

The objective is to extract the components of the matrix M in (3.8) using the least squares method as follows:

$$\mathbf{M} = \mathbf{U} \cdot \mathbf{A}^T \cdot (\mathbf{A}\mathbf{A}^T)^{-1} \quad (3.12)$$

Angular rate tests are used for calibrating scale factors and nonorthogonalities of the gyroscopes. If both rate tests and the improved 6-position static method are used together, one can determine all the inertial sensor errors for low-cost MEMS sensors.

3.3.2 Angular Rate Test

Angular rate tests are used for calibrating the biases, scale factors, and nonorthogonalities of gyroscopes for automotive-grade navigation systems. Rate tests are typically done using a precise rate turntable as demonstrated in Figure 3.11.

Figure 3.11 CIU unit on the turntable.

By rotating the unit through given turning rates and comparing the outputs of the IMU to these references, the biases, scale factors, and nonorthogonalities are estimated using the same method as explained in Section 3.3.1. This is typically accomplished by rotating the table through a defined angular rate in both clockwise and counterclockwise directions.

3.3.3 Thermal Calibration Test

There are two main approaches for thermal testing. The first allows the IMU enclosed in the thermal chamber to stabilize at a particular temperature corresponding to the temperature of the thermal chamber and then records the inertial sensor data. This method of recording the data at specific temperature points is called the *soak method*. The second method known as the *ramp method*, the IMU temperature is linearly increased or decreased for a certain period of time [8, 10]. After thermal testing, the sensors are calibrated for the temperature-dependent variations using simple modeling techniques. The thermal calibration is defined in the following example.

In order to compensate for the thermal drifts of low-cost MEMS sensors, thermal calibration may be conducted to estimate the biases and scale factor variations of the sensors over a range of temperatures. For this example, the temperature range is from –25°C to 70°C. In the test setup [1, 2] a turntable and a thermal chamber are assembled together to form a thermal turntable unit as shown in Figure 3.12.

Figure 3.12 Thermal test setup.

The rotating axis of the turntable is extended into the thermal chamber through a narrow round opening in the chamber's side wall. A low-cost MEMS IMU custom inertial unit (CIU) is placed in the chamber and is fixed on the extended tabletop of the turntable. This placement of the IMU allows the freedom of rotating it under controlled temperature. The data is collected in the Inertial Lab at the University of Calgary [1]. The IMU signals are sampled at 100 Hz at different temperatures and saved on a laptop via a 16-bit A/D card (DAQCard-6036E) from National Instrument for postprocessing. The calculated biases for accelerometers and gyroscopes at different temperatures by the soak method are shown in Figures 3.13 and 3.14, respectively.

For accelerometers, the bias drift can be as high as 1 m/s², while for gyroscopes the drift in biases can reach 5°/sec over the whole temperature range. Please note the constant bias estimated from the improved 6-position static test is first removed from each calculated bias at different temperature points. Furthermore, the biases for accelerometer and gyroscopes by the ramp method (with temperature gradient of 0.2°/sec) are shown in Figures 3.15 and 3.16, respectively.

As observed from Figures 3.13–3.16, the biases of accelerometers and gyroscopes vary significantly with temperature, and hence should be modeled to get accurate navigation results.

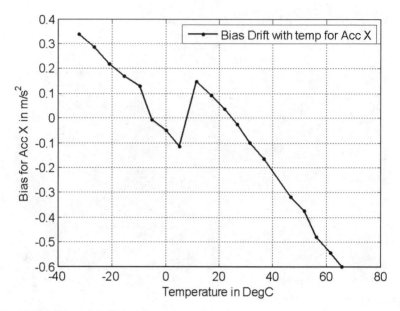

Figure 3.13   Thermal variation of accelerometer bias by the soak method.

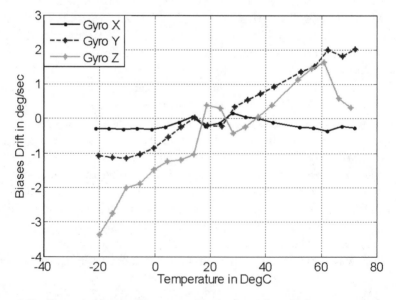

Figure 3.14   Thermal variation of gyroscope bias by the soak method.

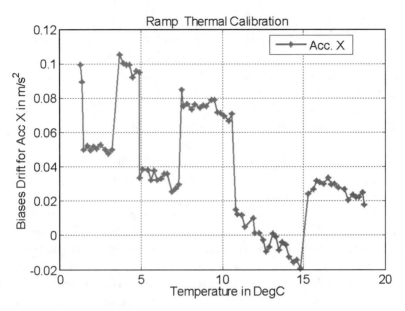

Figure 3.15 Thermal variation of accelerometer bias by the ramp method.

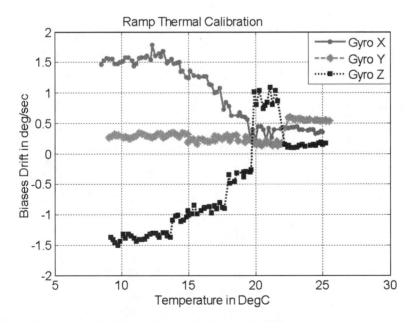

Figure 3.16 Thermal variation of gyroscope bias by the ramp method.

Figures 3.15 and 3.16 illustrate the biases and scale factor error drifts with temperature obtained from the ramp method. It is evident that these figures are very different from the ones obtained previously (Figures 3.13 and 3.14) by the soak method. This is because the temperature of the thermal chamber may not be the same as that of the IMU.

Similar to bias, the scale factor variation is also estimated for the soak and ramp methods. The scale factor error variation with temperature for accelerometer x and gyroscope z using the soak method is provided in Figure 3.17. After collecting the calibration parameters by different methods, thermal compensation models by simple linear interpolation methodology are obtained for each of the calibration methods.

In addition to the temperature-dependent correction parameters obtained by the soak and ramp methods, the sensor manufacturer also provides the temperature correction parameters. The manufacturer-provided parameters are different from the parameters obtained by the two lab tests. The objective of these thermal calibration and compensation tests is to establish which method is better, and if there is a need for these time-consuming calibrations by the soak and ramp methods or if manufacturer's specifications be used.

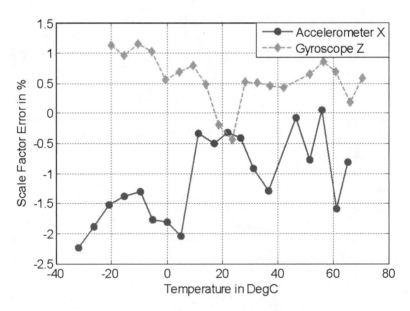

Figure 3.17 Thermal variation of accelerometer and gyroscope scale factor by the soak method.

Real field test data is used to evaluate the advantages and disadvantages of thermal compensation methods. Kinematic data collected in December 2005 by the Mobile Multi-Sensors research group (MMSS) is used for this analysis. The test data is collected with CIU, which is installed in the cargo area of a test vehicle. The test data is integrated with NovAtel Inc. OEM4 receiver single point positioning GPS (SPP GPS) solution. For the reference trajectory, a solution from a navigation-grade inertial system from Honeywell is integrated with differential GPS (DGPS) positions from the same OEM4 receiver.

The quality of the position estimation is often evaluated by simulating a set of short-term GPS signal outages where the inertial system works as a standalone navigation system. The IMU position errors during these short-term GPS signal outages are obtained by comparing the position errors to the reference trajectory solution. For example, in Figure 3.18 GPS signal outages of 30-seconds each are simulated in the field test dataset. This figure demonstrates the values of position errors obtained for all 4 GPS outages by the 6-position static test and the 2 lab temperature calibration tests (i.e., the soak and ramp methods).

It is clear that the position drift errors are the least for the soak compensation method followed by the 6-position static test method. For the ramp compensation method, the position drifts are significantly larger than the 6-position method.

Similarly, a second dataset is collected and analyzed, and results are obtained with longer GPS signal outages of 60-seconds each. Figure 3.19

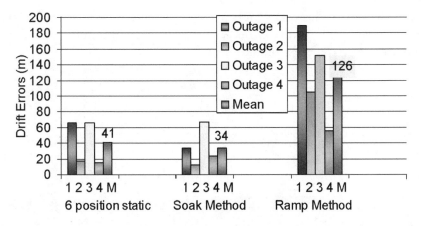

Figure 3.18 Position drift errors for CIU data by three methods.

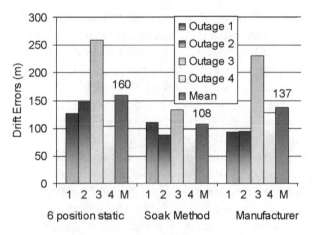

Figure 3.19 Drift errors for CIU data by three methods.

demonstrates the results for 6-position without any thermal calibration, soak method, ramp method, and the 6-position manufacturer-specified thermal correction values. The results from the ramp compensation method are over 500m and are not shown in the figure.

After compensating for the biases and scale factor errors by the soak, ramp, and manufacturer's datasheet values methods, we establish that the soak method is the best way to compensate the thermal variations effects. We further observe that the effect of thermal calibration is more evident when there is gradual variation in the temperatures of the collected data as compared to sudden gradual changes (for example, opening the door of the heated vehicle when the outside temperature is –30°/C).

Figure 3.20 illustrates typical deterministic and stochastic MEMS sensor errors. Note that if thermal errors are not calibrated and compensated, these errors can be classified as random errors.

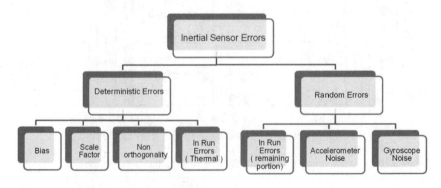

Figure 3.20 MEMS inertial sensor errors.

Random errors include the remaining in-run errors, accelerometer and gyroscope noises, residual systematic errors, and run-to-run variation errors. Since these error sources cannot be separated from the actual vehicle signal, these are best expressed by lumped parameters in the state vector.

3.4 Random/Stochastic Errors

Stochastic errors are the random errors that occur due to random variations of bias or scale factor over time, and are known as bias or scale factor drifts [1, 4, 12, 13]. The drift may also occur if other electronic equipment interferes with the output signals. The errors are random but can be modeled by stochastic processes. It is important to understand the differences among the stochastic processes to implement the most appropriate model for the dataset.

3.4.1 Examples of Random Processes

The basic difference between deterministic and stochastic modeling is that a deterministic modeling relationship has to be established between one or more inputs and one or more outputs, whereas in stochastic modeling, there may not be any direct relationship between the inputs and outputs [2]. Some common types of random processes are given in the following subsections that may be used to describe the random errors for the sensors.

3.4.1.1 Gaussian Random Process

A Gaussian random process $x(t)$ is one characterized by the property that joint probability distribution functions of all orders are multidimensional normal distributions [13]. For a Gaussian process, the distribution of $x(t)$ for any time t is the normal distribution, for which the density function is expressed by (3.13).

$$f_{X_t}(x_t) = \frac{1}{\sigma\sqrt{2\pi}} e^{-\frac{1}{2}(\frac{x-\mu}{\sigma})^2} \tag{3.13}$$

where μ is the mean and σ^2 is the variance of the Gaussian process. If $x(t)$ is an n-dimensional Gaussian vector then the distribution of $x(t)$ is the normal distribution expressed by (3.14).

$$f_{X_t}(x_t) = \frac{1}{(2\pi)^{n/2}\sigma} \exp[-\frac{1}{2}(x-\mu)^T \sigma^2 (x-\mu)] \tag{3.14}$$

All statistical properties of a Gaussian random process are defined by the first- and second-order moments of the distribution and are the easiest and most commonly used distribution [13].

3.4.1.2 Gauss-Markov Random Process

A stationary Gaussian process $x(t)$ that has an exponential autocorrelation is called a Gauss-Markov process [12, 13]. The autocorrelation and spectral functions are given by (3.15) and (3.16), respectively [14].

$$R_x(\tau) = \sigma^2 e^{-\beta|t|} \tag{3.15}$$

$$S_x(jw) = \frac{2\sigma^2\beta}{w^2 + \beta^2} \tag{3.16}$$

where β is the inverse of correlation time ($1/e$ point) as shown in Figure 3.21.

The spectral density function is obtained by taking the Fourier transform of the autocorrelation function as shown in (3.17)–(3.19).

$$S_x(jw) = \int_{-\infty}^{\infty} \sigma^2 e^{-\beta|\tau|} e^{-jw\tau} d\tau \tag{3.17}$$

$$S_x(jw) = \int_{-\infty}^{0} \sigma^2 e^{+\beta\tau} e^{-jw\tau} d\tau + \int_{0}^{\infty} \sigma^2 e^{-\beta\tau} e^{-jw\tau} d\tau \tag{3.18}$$

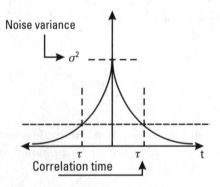

Figure 3.21 Autocorrelation function of Gauss-Markov process.

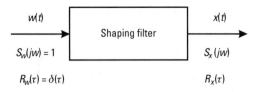

Figure 3.22 Shaping filter.

$$S_x(jw) = \int_{-\infty}^{0} \sigma^2 e^{+\beta\tau} e^{-jw\tau} d\tau + \int_{0}^{\infty} \sigma^2 e^{-\beta\tau} e^{-jw\tau} d\tau \qquad (3.19)$$

This process can be modeled by passing white noise through simple shaping filters as shown in Figure 3.22.

The system model for the shaping filter can be written by (3.20).

$$H(jw) \cdot H(-jw) = \frac{\sqrt{2\sigma^2\beta}}{\beta + jw} \frac{\sqrt{2\sigma^2\beta}}{\beta - jw} \qquad (3.20)$$

Taking the Laplace transform, we obtain (3.21).

$$H(s) \cdot H(-s) = \frac{\sqrt{2\sigma^2\beta}}{\beta + s} \frac{\sqrt{2\sigma^2\beta}}{\beta - s} \qquad (3.21)$$

The steady-state transfer function is then given by (3.22).

$$H(s) = \frac{\sqrt{2\sigma^2\beta}}{\beta + s} \qquad (3.22)$$

or

$$\frac{x(s)}{w(s)} = \frac{\sqrt{2\sigma^2\beta}}{\beta + s} \qquad (3.23)$$

$$\beta x(s) + s x(s) = \sqrt{2\sigma^2\beta} w(s) \qquad (3.24)$$

Now taking the inverse Laplace transform of (3.24), we obtain (3.25).

$$\dot{x}(t) = -\beta x(t) + \sqrt{2\sigma^2\beta}w(t) \tag{3.25}$$

where the spectral density is $2\sigma^2\beta$.

Generally, for most of the low-cost inertial systems (gyroscope drift 100–1,000°/h), the first-order Gauss-Markov model is utilized to describe random errors associated with inertial sensors. This Gauss-Markov model is represented by the system and measurement model of the following form.

$$x(k) = \Phi_{k-1}x(k-1) + G_{k-1}w(k-1) \tag{3.26}$$

$$z(k) = x(k) \tag{3.27}$$

where Φ_{k-1} is the state transition matrix, G_{k-1} is the shaping filter, $x(k)$ is the state vector, and $z(k)$ is the measurement vector.

The undefined matrices are obtained through the following steps:

1. Multiply (3.26) by $x(k-1)$ and take the expectation.

$$E[x(k)x(k-1)] = \Phi_{k-1}E[x(k-1)^2] + G_{k-1}E[w(k-1)x(k-1)] \tag{3.28}$$

As $E[w(k-1)x(k-1)] = 0$, $E[x(k-1)^2] = \sigma^2$, and $E[x(k)x(k-1)] = \sigma^2 e^{-\beta}$, we obtain (3.29).

$$\Phi_{k-1} = e^{-\beta} \tag{3.29}$$

2. Square each term and take the expectation of each term.

$$E[x(k)^2] = \Phi_{k-1}E[x(k-1)^2] + G_{k-1}E[w(k-1)^2] \tag{3.30}$$

On substituting known values and using $E[w(k-1)^2] = 0$, we obtain

$$\sigma^2 = \Phi_{k-1}^2\sigma^2 + G_{k-1}^2 \tag{3.31}$$

3. Substituting value of Φ_{k-1} from (3.29), we obtain

$$G_{k-1} = \sigma\sqrt{(1 - e^{-2\beta})} \tag{3.32}$$

The complete model becomes

$$x(k) = e^{-\beta}x(k-1) + \sigma\sqrt{(1 - e^{-2\beta})}w(k-1) \qquad (3.33)$$

where the covariance matrix is $\sigma^2(1 - e^{-2\beta})$.

This model is incorporated to define a process that exhibits a low correlation between values, which are well separated in time. If the estimation period of a process is smaller than the time constant of the Gauss-Markov process, the process can be modeled by the random walk (RW) process [12].

3.4.1.3 Random Walk Process

The RW process results when uncorrelated signals are integrated. For RW processes the difference between two immediate time epochs of a random variable is a purely random/white sequence, as given in (3.34).

$$\dot{b}(t) = w(t) \qquad (3.34)$$

where $\dot{b}(t)$ is a random variable with spectral density $E[w(t)w(\tau)] = q(t)\delta(t-\tau)$, $q(t)$ is the magnitude, and $w(t)$ is white noise. The RW process is obtained when the white noise process is integrated. The corresponding discrete time representation is given by (3.35).

$$b(k) - b(k-1) = w(k-1) \qquad (3.35)$$

with the noise covariance given by $q(k) = q(t_k - t_{k-1})$.

The RW took its name from considering an analogy with a person walking with a fixed step length (distance) in arbitrary directions. Note that RW is a nonstationary process as state uncertainty increases linearly with time. Gyroscope and accelerometer signals, which contain white noise components, are integrated to obtain angle and velocity. Hence angle and velocity values are corrupted with these integrated white noise components called angular random walk (ARW) and velocity random walk (VRW). ARW is the average deviation rate from the true value after the integration [12]. Similarly, for accelerometers the average deviation rate from the true value after integration of the acceleration is known as VRW. These effects need to be modeled as they are totally random in nature and cannot be removed by calibration.

3.5 Stochastic Modeling

Inertial sensor random errors are modeled by passing a white noise through shaping filters to yield an output of time-correlated noises. The values of the shaping filter parameters are estimated through minimization of the differences between the output of the shaping filter and the actual inertial sensor noise process. Two common methods of identifying and modeling the involved random processes are discussed in the following subsections.

3.5.1 Autocorrelation Function

In this method, the parameters are computed by evaluating the autocorrelation function of raw static data [13]. First, static data is collected for a long period of time. The longer the dataset, the better the process identification is, and more accurate parameter estimation occurs. The computation and plotting of the autocorrelation function will define the type of random process. Although seemingly straightforward, the procedure is often complicated by the fact that data may not be taken over a sufficient period of time and therefore other methods are generally used [13]. Another method for stochastic modeling is evaluating the power spectral density (PSD) of the data. The PSD is the Fourier transform of the autocorrelation function of the signal if the signal can be treated as a wide-sense stationary random process. The most common modeling technique to determine the characteristic of the underlying random inertial sensor noise for low-cost MEMS IMUs is the Allan variance [5] technique. It helps identify the source of a given noise term in the data.

3.5.2 Allan Variance Methodology

Allan variance is a method of representing root mean square (RMS) random drift error as a function of averaged time. Consequently, Allan variance method is incorporated to determine the characteristics of the underlying random processes that give rise to data noise [2, 5–7]. The data is normally plotted as the square root of the Allan variance versus T on a log-log plot. For very low-cost MEMS IMUs, we are mainly concerned with the high frequency noise terms that have a correlation time much shorter than the sample time and contribute towards the ARW, VRW, or low-frequency Gauss-Markov's model parameters. The Allan variance for VRW is given by (3.36).

$$\sigma(T) = Q/\sqrt{T} \tag{3.36}$$

where Q is the VRW coefficient, and each cluster has time T, which is equal to nt_0 as shown in Figure 3.23.

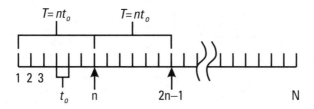

Figure 3.23 Data structure used by the Allan variance.

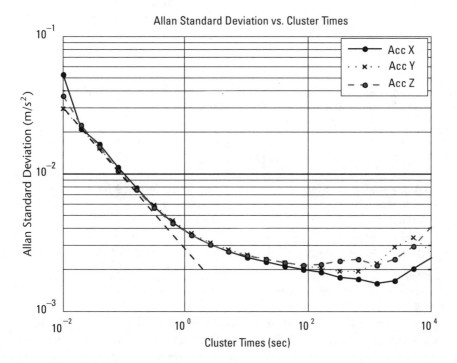

Figure 3.24 CIU accelerometer Allan variance results.

Equation 3.36 indicates that a log-log plot of $\sigma(T)$ versus T has a slope of $-1/2$. Furthermore, the numerical value of Q can be obtained directly by reading the slope line at $T = 1$ sec from Figures 3.24 and 32.5. A log-log plot of $\sigma(T)$ versus T provides a direct indication of the types of random processes, which exist in the inertial sensor data. The calculated VRW and ARW values are listed in Table 3.1.

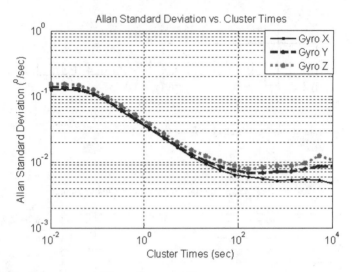

Figure 3.25   CIU gyroscope Allan variance results.

Table 3.1
Error Characteristics of CIU

Velocity Random Walk for Accelerometers $(m/s/\sqrt{hr})$				Angular Random Walk for Gyroscopes (deg/ $\sqrt{hr}$)			
Acc x	Acc y	Acc z	Average	Gyro x	Gyro y	Gyro z	Average
0.189	0.181	0.179	0.183	1.985	2.203	2.580	2.256

These parameters are required for designing the process noise matrix (Q) to be used in the extended Kalman filter algorithm in the GPS/INS integrated navigation systems. Details are provided in Chapters 6. At this point, we have defined all the possible error sources for the low-cost MEMS sensors. We now provide the complete error models that incorporate the errors defined earlier in Sections 3.2 and 3.4.

3.6 Sensors Measurement Models

Sensor measurement models, as the name indicates, are used to define different components of a measured signal. For example, every real sensor measurement consists of constant errors such as systematic and random errors. The following subsections discuss the details of sensor measurement models for accelerometers and gyroscopes.

3.6.1 Accelerometer Measurement Model

Accelerometers measure the specific force of the body but usually consist of unwanted systematic and nonsystematic signal components as shown from (3.37) [4].

$$I_f = f + b_{cons} + b_{random} + (S_{cons} + S_{random})f + S_2 f^2 + Nf + \delta g + \varepsilon(f) \quad (3.37)$$

where I_f is the actual measurement in (m/s^2); f is the true specific force (m/s^2); b_{cons} is the constant portion of the bias removed by calibration process (m/s^2); b_{random} is the remaining random portion of the bias modeled in the integrated INS/GPS algorithm (m/s^2); S_{cons} and S_{random} are the constant and random portions of the linear scale factor (ppm or %); S_2 is the nonlinear scale factor (ppm or %); N represents the nonorthogonalities (ppm or %); δg is the anomalous gravity (deviation from the theoretical gravity value) (m/s^2); and $\varepsilon(f)$ is the accelerometer sensor noise (m/s^2).

3.6.2 Gyroscope Measurement Model

A gyroscope is an angular rate sensor that provides either angular (rate sensing type) or attitude (rate integrating type) rates. The following model represents the single-axis gyroscope measurement of the angular rate.

$$I_w = w + b_{cons} + b_{run-to-run} + b_{in-run} + Sw + Nw + \varepsilon(w) \quad (3.38)$$

where I_w is the measurement (°/sec); w is the true angular velocity (°/sec); b_{cons} is the constant portion of the bias removed by the calibration process (°/sec); b_{random} is the remaining random portion of the bias modeled in the integrated INS/GPS algorithm (°/sec); S_{cons} and S_{random} are the constant and random portions of gyroscope scale factor (ppm or %); b_w is the gyroscope instrument bias (°/sec); S is the gyroscope scale factor; N is the nonorthogonalities of the gyroscope (ppm or %); and $\varepsilon(w)$ is the gyroscope sensor noise (°/sec).

We have defined the signal and provided the error removal or modeling techniques for low-cost MEMS sensors. Is this enough to start a meaningful navigation?

References

[1]   Aggarwal, P., et al., "A Standard Testing and Calibration Procedure for Low Cost MEMS Inertial Sensors and Units," *Journal of Navigation*, Vol. 61, No. 2, 2007, pp. 323–336.

[2]   Hou, H., and El-Sheimy, N., "Inertial Sensors Errors Modeling Using Allan Variance," *Proceedings of ION GNSS 2003*, Portland, Oregon, Sept. 9–12, 2003.

[3]   Hide, C.D., *Integration of GPS and Low Cost INS Measurements*, Ph.D. thesis, Institute of Engineering, Surveying and Space Geodesy, University of Nottingham, U.K., 2003.

[4]   El-Sheimy, N., "Inertial Techniques and INS/DGPS Integration," ENGO 623—course notes, Department of Geomatics Engineering, University of Calgary, Canada, 2006.

[5]   IEEE Std 952-1997, "IEEE Standard Specification Format Guide and Test Procedure for Single-Axis Interferometric Fiber Optic Gyros."

[6]   IEEE Std 1139-1988, "IEEE Standard Definitions of Physical Quantities for Fundamental Frequency and Time Metrology."

[7]   IEEE Std 1293-1998, "IEEE Standard Specification Format Guide and Test Procedure for Linear, Single-Axis, Non-gyroscopic Accelerometers."

[8]   Titterton, D., and Weston, J.L., *Strapdown Inertial Navigation Technology, Second Edition*, Bodmin, U.K.: American Institute of Aeronautics and Astronomy (AIAA), 2004, p. 558.

[9]   Walid, A.H., "Accuracy Enhancement of Integrated MEMS-IMU/GPS Systems for Land Vehicular Navigation Applications," Ph.D. Thesis, Department of Geomatics Engineering, University of Calgary, Canada, UCGE Report No. 20207, 2005.

[10]   Shcheglov, K., et al., "Temperature Dependent Characteristics of The JPL Silicon MEMS Gyroscope," *IEEE Aerospace Conference Proceedings*, Vol. 1, Mar 18–25, 2000.

[11]   Park, M., "Error Analysis and Stochastic Modeling of MEMS Based Inertial Sensors for Land Vehicle Navigation Applications," M.Sc. thesis, Department of Geomatics Engineering, University of Calgary, Canada, UCGE Report No. 20194, 2004.

[12]   Brown, R.G., and Hwang, Y.C., *Introduction to Random Signals and Applied Kalman Filtering, Second Edition*, New York: John Wiley & Sons Inc., 1992.

[13]   Gelb, A., *Applied Optimal Estimation*, Cambridge, MA: The Massachusetts Institute of Technology Press, 1974.

[14]   Grewal, A.P., and Andrews, M.S., *Kalman Filtering Theory & Practice Using MATLAB, Third Edition*, N.J.: Wiley-IEEE Press, 2008.

4

Initial Alignment of MEMS Sensors

4.1 Introduction

The procedure of determining the initial values of the attitude angles (pitch, roll, and azimuth) is known as alignment [1–3]. It is essential to align an inertial system before the start of navigation since any errors in the initial attitude angles may lead to relatively large position errors in the long term, thus deteriorating the overall performance and often leading to filter divergence. In general, the alignment procedure involves accelerometers and gyroscopes to monitor the specific forces and angular velocities along their respective orientation in space. Unless the inertial system is somehow aligned with the vehicle, the information provided by MEMS inertial sensors is not useful for navigating the vehicle [4]. Moreover, navigation is not possible if the IMU and the vehicle are perfectly aligned but the orientation of the IMU is unknown in some stable reference frame [5].

A complete flow chart of Chapter 4 is provided in Figure 4.1. To understand alignment, the reader must familiarize himself/herself with reference frames. Reference [1] is a good text to understand the reference frames, especially inertial frame, local level frame, and body frame.

Generally for alignment, it is assumed that the body frame (*b*-frame) of the vehicle and the inertial system are exactly the same and, therefore, the *b*-frame can be used interchangeably for the moving body or IMU frames as shown in the left panel of Figure 4.2.

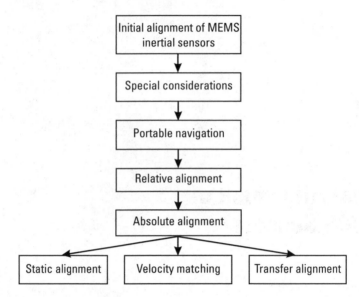

Figure 4.1   Flowchart for Chapter 4.

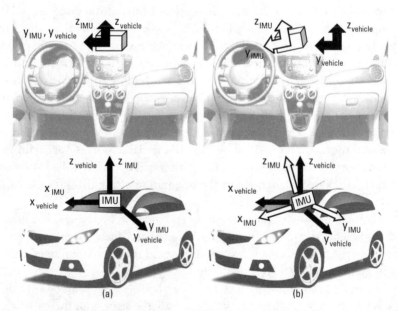

Figure 4.2   (a) shows a perfectly aligned IMU with the vehicle. (b) shows that the IMU is misaligned with respect to the vehicle frame.

There are times, however, when we have to distinguish between the *b*-frame of the IMU from the vehicle frame or *v*-frame. This is especially true for portable navigation systems in which a user is responsible for mounting the navigation system on the vehicle. In this situation, there is no guarantee that the IMU *b*-frame will coincide with the *v*-frame as shown in the right panel of Figure 4.2. MEMS sensors have opened the possibilities of such portable systems, and, therefore, we must look at the issues involved with their orientation.

When the IMU (*b*-frame) is aligned with the *v*-frame, the next step is to align the *b*-frame with the *l*-frame. This alignment and orientation is well developed and documented in literature [1, 2] and hence, the main focus of this chapter is to provide an overview of the different alignment methods, which can be used for low-cost MEMS sensors.

4.2 Considerations for MEMS Sensor Navigation

With the introduction of compact, low-power, and cost-efficient MEMS sensors, it is possible to have portable integrated INS/GPS navigation modules [7–9]. Recently, consumer navigation systems have become popular. GPS-based positions are displayed to the user on a digital map as shown in Figure 4.3. Despite their popularity, the recent systems suffer from reliability issues due to the GPS signal outages that can happen in urban areas. An easy solution to this problem is to integrate MEMS inertial sensors but the alignment of the unit becomes a critical issue.

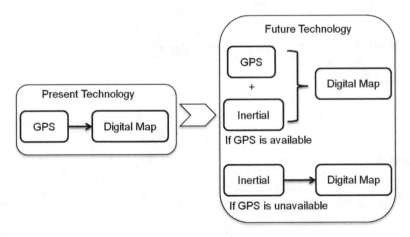

Figure 4.3 A possible future trend in portable navigation.

There are two types of alignment that are required before the navigation parameters can be estimated for a portable navigation module. The first alignment is the alignment of the IMU axes with respect to the vehicle axes (i.e., making *b*-frame coincide with *v*-frame), which is called relative alignment. Once relative alignment is achieved, the next step is to align the *b*-frame with the *l*-frame, which is called the absolute alignment. Refer to Figure 4.4 for a summary of the navigation process. There is a great deal of information available for *b*-frame with *l*-frame alignment and the reader is encouraged to consult [4–6]. To date, there is no discussion on the topic of *b*-frame and *v*-frame alignment (i.e., relative alignment), which is quite important for MEMS-based integrated INS/GPS navigation. In this chapter we are concentrating more on the relative alignment case.

4.3 Portable Navigation System

A portable navigation system is a system that can be carried by the user from a vehicle and be used for on-foot navigation needs. Recently, smart phones have provided such portability to the users using assisted GPS (AGPS). AGPS assists the GPS receiver to only look for the satellites that are visible accord-

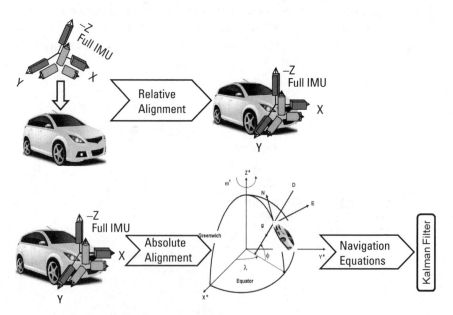

Figure 4.4 Summary of the navigation process.

ing to the geographic location estimated by the cell phone positioning. The message that contains the satellite visibility information is broadcast via cell phone signals. Such information is then used by the GPS receivers to try to find the weak satellite signals indoors to get a position fix. However, even with the help of cell towers, it is difficult for GPS signals to propagate in an indoor environment particularly in a deep indoor structure.

Therefore, a reliable navigation system cannot be totally dependent on GPS or AGPS. An inclusion of the MEMS inertial sensors is one of the ways to overcome the navigation interruptions due to the inability of signal propagation in certain environments. The inertial sensors require initialization, which consists of information about the absolute position, velocity, and orientation of the sensors with respect to a reference frame. Orientation of sensors refers to the alignment, and it is important to first orient the system perfectly with the moving body (i.e., the relative alignment) before the process of initialization can be started. There are two strategies that can be implemented to address the relative alignment in the design of a portable navigation system:

1. Start the navigation from a good GPS availability area and use the initial attitude errors from the navigation KF as the orientation parameters.

2. Restrict the orientation of the portable system to allowable limits.

There are advantages and disadvantages for both methods. Let us begin by analyzing the first strategy in detail. In this approach, the user does not need to orient the portable system and thus it is considered user friendly. However, to make the inertial sensors work, the system is required to stay stationary for a period of time to estimate roll, pitch, and initial absolute coordinates using GPS. Next, for heading determination, the system should be subjected to good dynamics under GPS lock. These restrictions are not realistic as it is quite likely that the system might be initialized in a vehicle parked in an underground parking facility. Therefore, it is sufficient to say that the condition of GPS availability may not be satisfactory for all common usage scenarios.

The second strategy involves limiting the orientation errors where the user is required to align the portable system such that the IMU *b*-frame and *v*-frame are in good agreement. The orientation accuracy solely depends on the user and it is safe to assume that even a careful user will be unable to properly align the system on each occasion. An easy way to solve this problem is the introduction of a holder inside the vehicle that is aligned with the vehicle and provides the user a feasible way to align the system for each use. This can be explained using the analogy of computer installation. Consider the task of attaching a traditional keyboard to a computer. Completing this task with ease is possible as there is only one port on the CPU that fits the keyboard cable's orientation. The introduction of a holder leads to an extra component required by the user. On the plus side, if such an aligned portable system is used, navigation can be started without the condition of a GPS signal lock.

4.4 Economical Considerations

Economically, partial MEMS inertial sensors are more attractive for portable navigation modules. A partial system consists of less than six sensors, which are three orthogonally oriented accelerometers and three orthogonal gyroscopes. Gyroscopes are usually more expensive than accelerometers, and, therefore, it is realistic to assume that gyroscope demand is lower from an economic perspective [10]. Keeping this in mind, a partial inertial module can be used for part of the navigation module, consisting only of a heading gyroscope and a dual- or triaxial accelerometer.

4.4.1 Economically Desirable Configuration

As mentioned before, for a MEMS-based inertial navigation module, a partial system is more desirable from an economic perspective. The relative alignment for the partial sensors, however, requires special attention as some of the information is missing. The following discussion involves options 1 and 2 given in Section 4.3 for the partial inertial system case to ensure that the reader understands the details involved in implementing such a system.

4.4.1.1 Known Parameters Design

Recall that the design of option 1 in Section 4.3 requires the availability of good GPS signals for the initiation of the inertial system. The initial roll and pitch values are estimated using the horizontal accelerometers output, whcih will be discussed in Section 4.5. After the initial orientation, this technique updates the attitude information using the Kalman filter; however, for a partial system, roll and pitch information cannot be observed on a continuous basis and with confidence due to the lack of sensors. Any change in the roll or pitch of the system during navigation (as presented in Figure 4.5) seriously degrades the navigation solution.

The only way to overcome this major problem is by implementing an algorithm that repeatedly looks for the change in orientation of the system and stops providing the heading information if a huge change is detected. Special

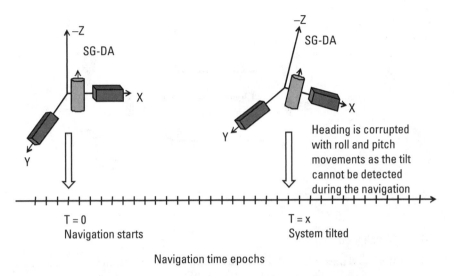

Figure 4.5 An obvious problem associated with known parameters design.

signal analysis is required in addition to the routine navigation algorithm to identify such problems. This is still an open research area and will not be discussed any further here.

4.4.1.2 Restrictive Design

The second option for the relative alignment is to simply restrict the system orientation. There are several methods that can be used to find the criteria for the restrictive design using partial inertial sensor configurations. The first method is more labor intensive but realistic as it requires the designer to collect the data for all possible misalignment scenarios. Afterwards, the designer has to establish the restrictions depending on the acceptable degradation threshold of the results.

Another method is to perform simulations using different misalignment scenarios. This method requires perfectly aligned full IMU data to ensure that only the simulated misalignment is affecting the results as shown in Figure 4.6. Misalignments are introduced to the aligned full IMU datasets by multiplying the signals with the direction cosines of the misalignment angles. Later roll and pitch gyro measurements are removed in the Kalman filter (KF) for processing. This method is purely simulation based yet it provides the results more quickly and is convenient to use for any number of trajectories. For example, due to the pitch angles, a trajectory that is collected on a level road is not the same as a trajectory collected on mountainous terrain.

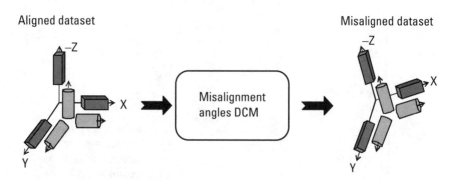

Figure 4.6 Misalignment simulations scheme.

4.4.1.3 Misaligned

Two different datasets are used in this example. After simulating a series of misalignments by multiplying the aligned data with the misalignment angles' DCM, both datasets are processed in partial sensor configurations in the KF. For a threshold of 20% degradation with respect to the aligned estimated attitude results, the acceptable misalignment for roll is determined to be ±15°. The process is presented in Figure 4.7 where the simulations for misalignment were conducted for ±45° of roll misalignments, and then the acceptable misalignment values were selected.

The threshold for degradation depends on the user's requirement. It could be less than 20% depending on the required accuracies of the attitude. However, for this example, a 20% degradation was selected to show the procedure.

Similarly, pitch misalignment design criteria can be determined. In this case, the pitch angle is misaligned and the attitude angles are estimated inside a KF for partial sensors (i.e., the KF did not use the data from the roll and pitch gyroscopes). The results are provided in Figure 4.8. It is important to note that higher roll and pitch values result in higher heading error due to the lack of roll and pitch sensors. The KF is accumulating these roll and pitch errors in the heading. This point can be verified by examining the heading misalignment case, which is discussed next.

Roll and pitch misalignments need to be restricted to ±15° as they start to degrade the heading of the navigation solution, which is the most important part. The heading misalignment is also studied by simulations and is provided in Figure 4.9. The results indicate that misalignments in the heading are captured by the KF solution more accurately due to the presence of the heading gyroscope.

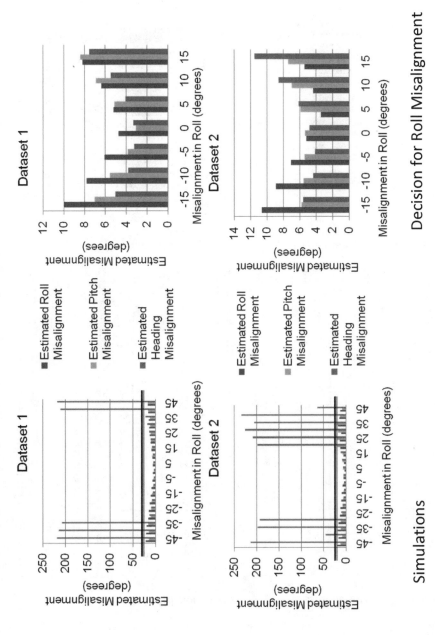

Figure 4.7 Procedure to obtain roll design restrictions.

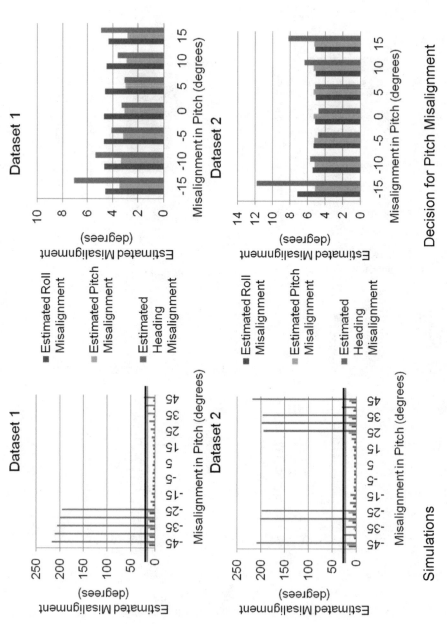

Figure 4.8   Procedure to obtain pitch design restrictions.

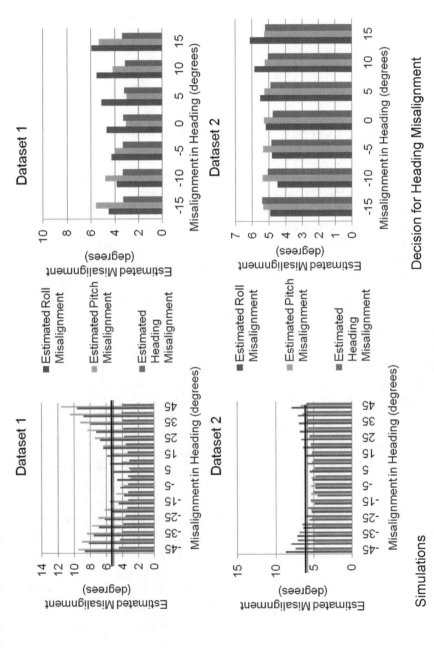

Figure 4.9 Procedure to obtain heading design restrictions.

The position drift errors for the two datasets are also provided in Figure 4.10. Both the full range and the zoomed-in portion for the misalignment range of ±15° are shown in this figure. This example only provides an illustration of the design of a system that contains a certain number of sensors, which is less than six. This method can be used for any number of sensors for restrictive design criteria determination.

4.4.2 Complete Six DOF IMU—Economically Less Desirable

Now we explore the relative alignment options for a full IMU or a complete six degrees of freedom (DOF) IMU. This system has all the sensors, and, in theory, it has the capability to measure the misalignments in roll, pitch, and heading attitudes. The known parameter design includes the initial alignment of the system using the horizontal accelerometers. The heading is computed as soon as the vehicle starts to move under good GPS availability. Also, during the movement of the vehicle, if the IMU is misaligned this change is detected by the outputs of roll and pitch gyroscopes in the KF.

If a full IMU is present, the restrictive design is not required as the orientation error is measured by the respective gyroscope. An example is provided below, which shows that the introduction of any errors in alignment whether it is in roll, pitch, or heading, does not affect the accuracy of results. This confirms the results when the SG-DA or SG-TA configuration is used to estimate the simulated heading misalignments. The resultant attitude errors are small and are correctly identified to be in the heading.

4.4.2.1 Relative Alignment Example

Similar datasets are used to study the design criteria and the importance of relative alignment for a full IMU. Figure 4.11 shows the capability of the KF to estimate the misalignment in the appropriate attitude. The first graph shows the results of the KF when simulated misaligned roll angles are introduced as part of the input. The KF successfully estimated the roll misalignments by utilizing the roll gyroscope data and provided the same results for the pitch and heading misalignment cases. The last graph shows the typical increase in drift errors for the simulated misalignments in the attitude angles. It is important to note that the maximum degradation in drift errors was close to 12%. These results indicate that if the full six DOF IMU is available, restrictive design is not necessary.

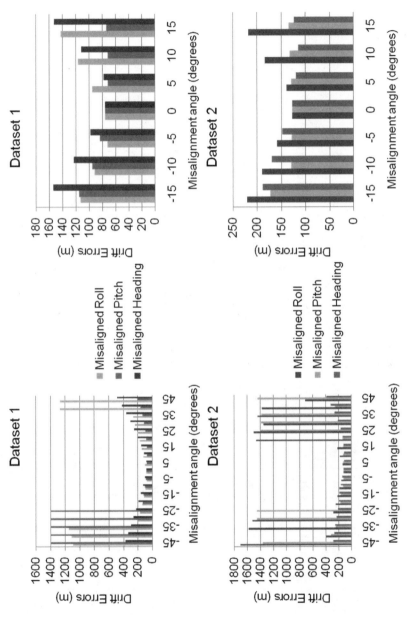

Figure 4.10 Position drift errors for the allowed misalignment range.

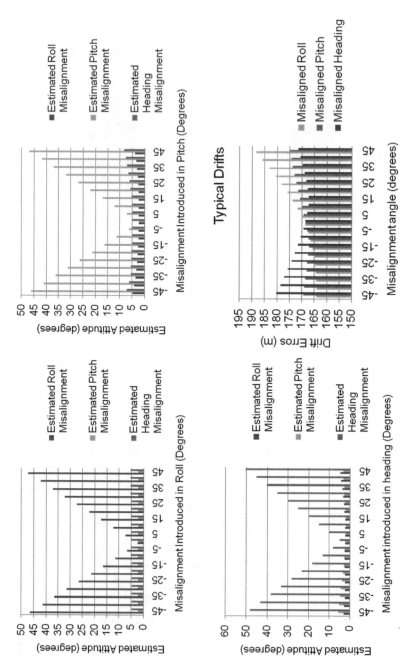

Figure 4.11 Importance of relative alignment for full IMU.

4.5 Absolute Alignment

Upon completion of relative alignment, the inertial systems require absolute alignment whereby the orientation of the axes of an inertial system is determined with respect to a reference frame such as an l-frame.

4.5.1 Static Alignment for MEMS Sensors

For MEMS sensors, the strong gravity signals from accelerometers can be measured and the accelerometer leveling is used to determine the angular displacements in the roll and pitch of the inertial system with respect to the l-frame. In summary, the accelerometer leveling is aligning the z-axis of the accelerometer triad (z^b) to the z-axis of the local level frame or l-frame (z^l) by driving the output of the horizontal accelerometers to 0. Equations (4.1) and (4.2) provide the mathematical formula to determine the roll and pitch of an orthogonal and stationary accelerometer triad. The details can be found in [1].

$$r = sign(f_z)\sin^{-1}(f_y/g) \tag{4.1}$$

$$p = -sign(f_z)\sin^{-1}(f_x/g) \tag{4.2}$$

where r is the roll angle;

p is the pitch angle;

$f_x, f_y,$ and f_z are the accelerometer signals; and

g is the gravity.

To initialize the navigation KF, all three attitude angles including roll, pitch, and heading (azimuth) are required. The accelerometer leveling can only provide the roll and pitch angles. For azimuth, usually a method known as gyroscope compassing is used. However, MEMS sensors have significantly high drift rates and noise characteristics, and, therefore, the gyroscope outputs cannot be used to estimate the azimuth or heading of the vehicle. The main reason is that gyroscope compassing uses the rotation rate of the Earth, which cannot be monitored with MEMS sensors. The heading is also important for the initialization of the navigation algorithm and, hence, requires some other method.

4.5.2 Static Alignment Example

Consider the example in which a vehicle started from a static period. The static period is approximately 300 seconds and only the last 10 points of the static periods are shown in Figure 4.12. The north and east accelerometer outputs of an actual dataset are shown. The first set corresponds to the situation when the inertial system is almost well aligned with the *l*-frame. In the second case, a misalignment of 5° is introduced in the east channel for comparisons. For leveling, the north and east sensing accelerometers outputs are averaged for the entire static period for a better computation of roll and pitch angles.

First the static portions of the outputs are averaged for both aligned and misaligned sets as shown in Table 4.1.

Next, using the formula for pitch and roll as provided in (4.1) and (4.2) for the NED frame, the two attitude angles are computed in radians. Since it is easier to visualize in decimal degrees, the radians are then converted to degrees by multiplying the results with 180° and dividing it with π. The computed values are given in Table 4.2.

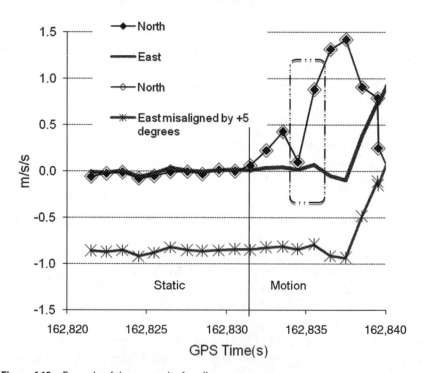

Figure 4.12 Example of the necessity for alignment.

Table 4.1
Aligned and Misaligned Values

Aligned	North (f_x)	East (f_y)	Misaligned	North (f_x)	East (f_y)
Average (m/s²)	0.011698	−0.008254	Average (m/s²)	0.011698	−0.865718

Table 4.2
Aligned and Misaligned Values

Aligned Values	
Pitch	**Roll**
$p = \sin^{-1}(0.012/9.81)$ $p = 0.068°$	$r = -\sin^{-1}(-0.008/9.81)$ $r = 0.048°$
Misaligned Values	
Pitch	**Roll**
$p = \sin^{-1}(0.012/9.81)$ $p = 0.068°$	$r = -\sin^{-1}(-0.866/9.81)$ $r = 5.060°$

This describes the procedure of accelerometer leveling. For high-grade inertial systems, a similar procedure can be used by utilizing the Earth's rotation rate. Such a procedure is not possible for the MEMS-grade sensors.

4.6 Velocity Matching Alignment

The GPS positions and velocity measurements can be used for the alignment of the IMU. Heading or the azimuth of a vehicle is determined by incorporating north and east velocity components from the GPS receiver. Along with the roll and pitch information calculated by using the accelerometer signals, the vehicle's attitude is estimated by incorporating the GPS-derived velocities. At every GPS update, the positions, velocities, and heading are updated to improve the navigation solution accuracy. Figure 4.13 shows the *l*-frame velocity vector and its relationship to the heading angle (A).

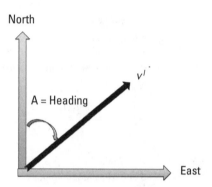

Figure 4.13   Geometrical relationship between heading and *l*-frame velocity.

The heading is always measured from the north direction, and, therefore, is written in terms of (4.3). However, care must be taken if there is any lever arm between the GPS receiver and the IMU. The lever arm must be compensated for before a GPS position update or attitude determination.

$$A = \tan^{-1}(V^E / V^N) \tag{4.3}$$

Usually, a KF is used to estimate the navigation parameters. The velocity from the GPS is fed back into the inertial navigation algorithm commonly known as a mechanization equation.

4.6.1   GPS Derived Heading Example

Referring to Figure 4.12, the heading was computed using the data points given in the grey shaded area. From this point, the heading of the vehicle was 86.2°.

4.7   Transfer Alignment

One shot alignment is typically performed if the navigation parameter from a high-accuracy source is available. Here, the attitude parameters are copied as the a priori information at the start of navigation for the low-cost MEMS IMU. The navigation KF in the case of a priori information starts from those parameters and then gets updated once the measurements become available. The navigation KF further improves the attitude.

The alignment procedures are well documented and readers are encouraged to read [1] for details. The main purpose of this book is to provide the user with an idea of what methods can be used to align a MEMS IMU. There are other methods as well but they are not possible for the highly noisy MEMS sensors. Among those methods, one of the common alignments is to perform analytic coarse alignment followed by a fine alignment. The details are in [1], but cannot be used here as it uses Earth's rotation rate to estimate the heading. Fine alignment is the introduction of a KF to refine the attitude obtained from the coarse alignment, and, again, it cannot be used due to the high error characteristics of the MEMS sensors.

Once these viable inertia sensor measurements are available to us, how do we estimate our position? Chapter 5 covers the navigation equations that link inertial sensor measurements to the object's position.

References

[1]   Titterton, D., and Weston, J.L., *Strapdown Inertial Navigation Technology, Second Edition*, Bodmin, U.K.: American Institute of Aeronautics and Astronomy (AIAA), 2007, p. 558.

[2]   Grewal, M.S., Weill, L.R., and Andrews, A.P., *Global Positioning Systems, Inertial Navigation, and Integration*, New York: John Wiley & Sons, Inc., 2001.

[3]   Savage, P., *Strapdown Analytics*, Parts 1 & 2, Maple Plain, MN: Strapdown Associates, 2000.

[4]   Syed, Z.F., et al., "Civilian Vehicle Navigation: Required Alignment of the Inertial Sensors for Acceptable Navigation Accuracies," *IEEE Transactions on Vehicular Technology*, Vol. 57, No. 6, 2008, pp. 3402–3412.

[5]   Wang, X., and Shen, G., "A Fast and Accurate Initial Alignment Method for Strapdown Inertial Navigation System on Stationary Base," *J. Control Theory Appl.*, Vol. 3, No. 2, May 2005, pp.145–149.

[6]   Shin, E.-H., *Estimation Techniques for Low-Cost Inertial Navigation*, PhD. thesis, May 2005, Department of Geomatics Engineering, University of Calgary, Canada, UCGE Report 20219, 2005.

[7]   "TomTom, Portable GPS Car Navigation Systems," http://www.tomtom.com, visited on October 3, 2007.

[8]   "Sony: Personal Navigation," http://www.sony.co.uk/view/ShowProductCategory .action?site=odw_en_GB&category=ICN+Personal+navigation, visited on October 3, 2007.

[9]   "Garmin. We'll Take You There," http://www.garmin.com/garmin/cms/site/us, visited on October 3, 2007.

[10] "Analog Devices Inc.," www.analog.com, visited November 19, 2009.

5

Navigation Equations

In Chapters 1–4, MEMS inertial sensors were introduced in detail. General applications for MEMS sensors were discussed, and specific usage of these sensors in the navigation field were provided. It was also shown that inertial navigation cannot be initiated before removing the systematic errors and developing an appropriate stochastic model for these very low-cost MEMS sensors. Any residual systematic error will contribute towards large drift errors during the integration of sensor measurements. Similarly, an unsuitable stochastic error model adds to the errors by corrupting the states prediction in the Kalman filter (KF) (which is discussed in the Chapter 6). Specific methods for alignment were also provided in Chapter 4, which is the essential step towards the implementation of the navigation algorithm. Alignment is a procedure to initialize gyroscope measurements, usually in the l-frame.

Chapters 5 focuses on the implementation of the navigation equations to begin relative navigation of the body in the defined reference frame. The navigation equations provide a numerical tool to implement the physical phenomena that relates the inertial sensor measurements to the navigation state (i.e., position, velocity, and attitude). The inertial sensor measures the body accelerations and angular rotations along three mutually orthogonal directions with respect to the body frame (b-frame). However, these measurements need to be converted into a well-defined reference frame for navigation. The reference frame is usually either a local level frame (l-frame) or an Earth fixed frame (e-frame).

There are two different methods to solve the navigation equations: the classical method and the half-interval method. The classical method is computationally less expensive and is easier to implement, but its usage is limited to very low-dynamic applications. On the contrary, the half-interval method is

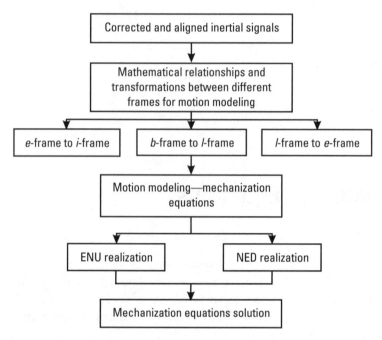

Figure 5.1   Chapter flowchart.

more complicated but provides a robust solution for high-dynamic navigation applications. In this book, we do not delve into the details of either the classical or the half-interval method, and we suggest the reader seek details in [1, 7–9]. A complete flowchart of this chapter is provided in Figure 5.1.

5.1   Introduction—Mathematical Relations and Transformations Between Frames

It is important to define the relationship between different frames before the derivation of navigation equations as this knowledge is crucial to obtain navigation solutions. Three different frame transformations are provided here [1].

5.1.1   *e*-Frame to *i*-Frame

The *e*-frame is referred to as the Earth fixed frame, which is fixed at the center of the Earth, while the *i*-frame is defined as the inertial frame [1]. The *z*-axis of the *e*-frame and *i*-frame are parallel to each other but the *e*-frame is rotating at an angular velocity with respect to the *i*-frame. This results in the Earth rotation ω_{ie}^e of the *e*-frame being expressed as:

$$\omega_{ie}^e = (0, 0, \omega^e)^T \tag{5.1}$$

Therefore, the direction cosine matrix (DCM) between these two frames is a single rotation around the *z*-axis as illustrated by (5.2). The angle of rotation is determined by taking a product of the ω_{ie}^e (in short ω^e) with the elapsed time (*t*).

$$R_e^i = \begin{pmatrix} \cos \omega^e t & -\sin \omega^e t & 0 \\ \sin \omega^e t & \cos \omega^e t & 0 \\ 0 & 0 & 1 \end{pmatrix} \tag{5.2}$$

5.1.2 ENU *l*-Frame to *e*-Frame

The DCM from ENU *l*-frame to *e*-frame can be expressed in terms of two rotations. The first rotation is around the *x*-axis (R_1) to align the vertical axes of the two frames. Once the verticals are aligned, the horizontal axes are aligned by a rotation around the *z*-axis (R_3). The R_1 and R_3 rotations for an arbitrary angle θ are written as follows:

$$R_1(\theta) = \begin{pmatrix} 1 & 0 & 0 \\ 0 & \cos \theta_1 & \sin \theta_1 \\ 0 & -\sin \theta_1 & \cos \theta_1 \end{pmatrix} \tag{5.3}$$

and

$$R_3(\theta) = \begin{pmatrix} \cos \theta_2 & \sin \theta_2 & 0 \\ -\sin \theta_2 & \cos \theta_2 & 0 \\ 0 & 0 & 1 \end{pmatrix} \tag{5.4}$$

In the transformation from the *l*-frame to the *e*-frame, the *e*-frame is taken as the reference and the *l*-frame is rotated. The following steps transform the ENU *l*-frame to an *e*-frame.

1. The first step is to rotate around the x-axis by an angle equal to $\phi - 90$ (where ϕ is the latitude) to bring the position in the horizontal plane. Here, θ_1 in (5.3) is replaced by $\phi - 90$. Performing this rotation makes the l-frame's z-axis or vertical axis parallel to the z-axis of the Earth.

2. Next, the newly formed z-axis is rotated back by the longitude angle (λ) and then by 90° to make the x-axes and y-axes of the new system parallel to the Earth reference frame. Here, θ_2 in (5.4) is replaced by $-90 - \lambda$. Combining the above two transformations gives us the required DCM:

$$R_\ell^e = R_3(-90 - \lambda) \cdot R_1(\phi - 90) = \begin{pmatrix} -\sin\lambda & -\sin\phi\cos\lambda & \cos\phi\cos\lambda \\ \cos\lambda & -\sin\phi\sin\lambda & \cos\phi\sin\lambda \\ 0 & \cos\phi & \sin\phi \end{pmatrix} \quad (5.5)$$

For a typical location on the Earth's surface, the angular velocity component (ω^e) is a function of the latitude as shown in Figure 5.2. Using simple geometry, the angular velocity vector in the l-frame, due to Earth's rotation, is:

$$\omega_{ie}^\ell = \begin{pmatrix} 0 & \omega^e \cos\varphi & \omega^e \sin\varphi \end{pmatrix}^T \quad (5.6)$$

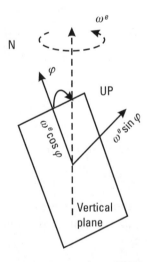

Figure 5.2   Components of Earth's rotation rate in an ENU frame.

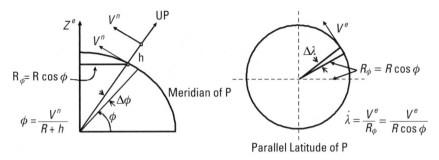

Figure 5.3   Components of the Earth's rotation rate in an ENU frame.

For a moving platform, the *l*-frame changes its position, and this causes a rotation of the *l*-frame with respect to the *e*-frame. The rates can be given in terms of the changes in latitude and longitude. When this rotation is given in the *l*-frame, it is commonly known as the transport rate as depicted by (5.7).

$$\omega_{el}^{\ell} = \begin{pmatrix} -\dot{\phi} & \dot{\lambda}\cos\phi & \dot{\lambda}\sin\phi \end{pmatrix}^{T} \tag{5.7}$$

This equation is defined on the basis of the following example. Let's consider a vehicle that is moving with east and north velocity components of V^e and V^n, respectively. Once the vehicle comes to a point *P* on the surface of the Earth, its relative components are determined using the relationship in Figure 5.3.

5.1.3   NED *l*-Frame to *e*-Frame

To rotate the *e*-frame onto the NED *l*-frame, the first required rotation is around the *z*-axis with the longitude angle. This takes care of the horizontal displacement of the two axes. The second rotation is around the new *y*-axis to take care of the latitude difference. These two rotations make all the axes of the rotated system parallel to the *e*-frame as given by (5.8).

$$R_e^l = R_2\left(-\varphi - \pi/2\right)R_3\left(\lambda\right) = \begin{bmatrix} -\sin\varphi\cos\lambda & -\sin\varphi\sin\lambda & \cos\varphi \\ -\sin\lambda & \cos\lambda & 0 \\ -\cos\varphi\cos\lambda & -\cos\varphi\sin\lambda & -\sin\varphi \end{bmatrix} \tag{5.8}$$

The l-frame to e-frame are simply the transpose of the above matrix. The angular velocity vectors describing the rotation of Earth in the l-frame, and the rotation of the l-frame with respect to the Earth fixed frame in the NED frame are given by (5.9) and (5.10).

$$\omega_{ie}^{\ell} = \left(\omega^e \cos\varphi \quad 0 \quad \omega^e \sin\varphi\right)^T \tag{5.9}$$

$$\omega_{e\ell}^{\ell} = \left(\dot{\lambda}\cos\phi \quad -\dot{\phi} \quad -\dot{\lambda}\sin\phi\right)^T \tag{5.10}$$

5.1.4 *b*-Frame to ENU *l*-Frame

If the axes of the body frame are perfectly aligned with the axes of the l-frame, the roll, pitch, and azimuth angles are all zeros and no transformation is required. Generally, this doesn't occur and rotations are required to transform the b-frame to an l-frame. For azimuth, the positive rotation is from north to east. We start by rotating the frames with the azimuth angle around the z-axis. The next rotation corrects for the pitch around the x-axis and finally the roll angle is corrected, which is the rotation around the y-axis.

$$R_l^b = R_2(r)R_1(p)R_3(-A) \tag{5.11}$$

R_1 and R_3 rotation matrices have been provided before in Section 5.1.2; R_2 rotation for an arbitrary angle θ is given as follows:

$$R_2(\theta) = \begin{pmatrix} \cos\theta & 0 & -\sin\theta \\ 0 & 1 & 0 \\ \sin\theta & 0 & \cos\theta \end{pmatrix} \tag{5.12}$$

The most expanded form of (5.11) is provided in (5.13).

$$R_l^b =$$

$$\begin{bmatrix} \cos A \cos r + \sin A \sin p \sin r & -\sin A \cos r + \cos A \sin p \sin r & -\cos p \sin r \\ -\sin A \cos p & \cos A \cos p & \sin p \\ -\cos A \sin r - \sin A \sin p \cos r & -\sin A \sin r - \cos A \sin p \cos & \cos p \cos r \end{bmatrix} \tag{5.13}$$

Using the properties of orthogonality, the rotation from the b-frame to l-frame can be done by simply taking a transpose of R_l^b. Hence, $R_b^l = R_l^{bT}$ and is provided in (5.14).

$$R_b^l =$$

$$\begin{bmatrix} \cos A \cos r + \sin A \sin p \sin r & -\sin A \cos p & \cos A \sin r - \sin A \sin p \cos r \\ -\sin A \cos r + \cos A \sin p \sin r & \cos A \cos p & -\sin A \sin r - \cos A \sin p \cos r \\ -\cos p \sin r & \sin p & \cos p \cos r \end{bmatrix} \quad (5.14)$$

The angular velocity vector between the two frames is expressed as the rate of change of the attitude angles.

$$\omega_{b\ell}^b = \begin{pmatrix} \dot{p} & \dot{r} & \dot{A} \end{pmatrix}^T \quad (5.15)$$

The Euler angles are the essential part of the navigation as the accelerations recorded in the *b*-frame needs to be converted into the reference frame using these angles.

5.1.5 *b*-Frame to NED *l*-Frame

A similar procedure is used to rotate the NED *l*-frame onto the *b*-frame. The *b*-frame is rotated based on the Euler angles by first correcting for roll, then pitch, and finally for azimuth.

$$R_b^l = R_3(-A)R_2(-p)R_1(-r) \quad (5.16)$$

$$R_b^l =$$

$$\begin{bmatrix} \cos r \cos A & -\cos p \sin A + \sin p \sin r \cos A & \sin p \sin A + \cos p \sin r \cos A \\ \cos r \sin A & \cos p \cos A + \sin p \sin r \sin A & -\sin p \cos A + \cos p \sin r \sin A \\ -\sin r & \sin p \cos r & \cos p \cos r \end{bmatrix} \quad (5.17)$$

Consequently, the angular velocity vector between the two frames is expressed as:

$$\omega_{b\ell}^b = \begin{pmatrix} \dot{r} & \dot{p} & \dot{A} \end{pmatrix}^T \quad (5.18)$$

5.2 Motion Modeling in the *l*-Frame

The *l*-frame formulation has the advantage that its axes are aligned to the local east, north, and vertical directions as compared to the *e*-frame. Therefore, the attitude angles (i.e., pitch, roll, and azimuth) are directly obtained as output of the navigation equations. Further, as the definition of the *l*-frame is based on the perpendicular to the reference ellipsoid, the geodetic coordinate differences $\Delta\phi$, $\Delta\lambda$, Δh are the outputs of the system. In addition, the computational errors in the navigation parameters on the horizontal plane (northeast plane) are bounded due to Schuler effect. This effect stipulates that the INS velocity and attitude errors along the horizontal plane are coupled together producing what is called the Schuler loop. This includes the strong coupling between the north velocity and the roll angle as well as between the east velocity and the pitch angle. As a result, these errors are bounded over time and oscillate with the Schuler frequency (1/5,000 Hz).

5.2.1 ENU Realization

The position of a moving platform in the *l*-frame is expressed in terms of the curvilinear coordinates (ϕ, λ, h).

$$r^\ell = \begin{pmatrix} \varphi & \lambda & h \end{pmatrix}^T \qquad (5.19)$$

The velocity is expressed by three components along the east (V^e), north (V^n), and vertical up direction (V^u).

$$V^\ell = \begin{pmatrix} V^e & V^n & V^u \end{pmatrix}^T \qquad (5.20)$$

The position is commonly represented in the curvilinear coordinate system as shown above, while the velocity is given in the Cartesian coordinate system. A rate of change in position, called the *velocity*, requires geometrical consideration to account for the two different coordinate systems. Therefore, the time rate of change of the position components is related to the velocity components as follows [2–4].

$$\dot\varphi = \frac{V^n}{M + h} \qquad (5.21)$$

$$\dot{\lambda} = \frac{V^e}{\left(N + h\right)\cos\varphi} \tag{5.22}$$

$$\dot{h} = V^u \tag{5.23}$$

Equations (5.21)–(5.23) are usually written in matrix notation as shown in (5.24).

$$\dot{r}^\ell = \begin{pmatrix} \dot{\varphi} \\ \dot{\lambda} \\ \dot{h} \end{pmatrix} = \begin{pmatrix} 0 & \dfrac{1}{M+h} & 0 \\ \dfrac{1}{(N+h)\cos\varphi} & 0 & 0 \\ 0 & 0 & 1 \end{pmatrix} \begin{pmatrix} V^e \\ V^n \\ V^u \end{pmatrix} = D^{-1}V^\ell \tag{5.24}$$

The M and N are radiuses of the curvature of the Earth in the meridian and prime vertical directions. Equation (5.24) is known as the position mechanization equation and requires computed quantities, which include orthogonal velocity components and the latitude.

Velocity mechanization equations use the measured accelerometers signals. Since the measured signals are in the b-frame, the first step is to convert them into the l-frame.

$$f^\ell = \begin{pmatrix} f^e & f^n & f^u \end{pmatrix}^T = R_b^\ell f^b = R_b^\ell \begin{pmatrix} f_x & f_y & f_z \end{pmatrix}^T \tag{5.25}$$

However, the acceleration components expressed in the l-frame, f^l, cannot directly provide the l-frame velocity components of the moving platform for three reasons. The first reason is the rotation rate ($\omega^e = 15°/hr$) of the Earth as given in (5.6). The change of orientation of the l-frame with respect to the e-frame due to the definition of the local north and up directions is the second reason. The north direction is tangential to the meridian all the time, while the up direction is normal to the Earth's surface. This effect is expressed by the angular velocity vector given in (5.7).

Substituting values of $\dot{\phi}$ and $\dot{\lambda}$ into (5.7) yields (5.26).

$$w_{el}^{\ell} = \begin{pmatrix} -\dot{\varphi} \\ \dot{\lambda}\cos\varphi \\ \dot{\lambda}\sin\varphi \end{pmatrix} = \begin{pmatrix} -\dfrac{Vn}{M+h} \\ \dfrac{V^e}{N+h} \\ \dfrac{V^e\tan\varphi}{N+h} \end{pmatrix} \tag{5.26}$$

The third reason is the normal gravity vector, which is a function of latitude and ellipsoidal height and is a dominant factor in the l-frame velocity mechanization [5].

$$g = a_1(1 + a_2\sin^2\varphi + a_3\sin^4\varphi) + (a_4 + a_5\sin^2\varphi)h + a_6h^2 \tag{5.27}$$

where

a_1 = 9.7803267715 m/sec^2;

a_2 = 0.0052790414;

a_3 = 0.0000232718;

a_4 = −0.000003087691089 1/sec^2;

a_5 = 0.000000004397731 1/sec^2; and

a_6 = 0.000000000000721 1/(m sec^2).

The l-frame gravity vector is written as:

$$g^{\ell} = (0 \quad 0 \quad -g)^T \tag{5.28}$$

Considering the above factors, the l-frame velocity component is expressed in (5.29).

$$\dot{V}^{\ell} = R_b^{\ell} f^b - (2\Omega_{ie}^{\ell} + \Omega_{e\ell}^{\ell})V^{\ell} + g^{\ell} \tag{5.29}$$

where Ω_{ie}^{ℓ} and $\Omega_{e\ell}^{\ell}$ are the skew-symmetric matrices corresponding to ω_{ie}^{ℓ} and $\omega_{e\ell}^{\ell}$, respectively, and are expressed as follows [8, 9]:

$$\Omega_{ie}^{\ell} = \begin{pmatrix} 0 & -\omega^e \sin\varphi & \omega^e \cos\varphi \\ \omega^e \sin\varphi & 0 & 0 \\ -\omega^e \cos\varphi & 0 & 0 \end{pmatrix} \qquad (5.30)$$

$$\Omega_{e\ell}^{\ell} = \begin{pmatrix} 0 & \dfrac{-V^e \tan\varphi}{N+h} & \dfrac{V^e}{N+h} \\ \dfrac{V^e \tan\varphi}{N+h} & 0 & \dfrac{V^n}{M+h} \\ \dfrac{-V^e}{N+h} & \dfrac{-V^n}{M+h} & 0 \end{pmatrix} \qquad (5.31)$$

The final part of the mechanization is the equation for attitude. The attitude angles of a moving rigid body are determined by solving the time derivative equation of the transformation matrix R_b^{ℓ}. Since the mechanization is implemented in the *l*-frame, the following time-derivative transformation matrix equation should be considered:

$$\dot{R}_b^{\ell} = R_b^{\ell}\Omega_{\ell b}^b \qquad (5.32)$$

The angular velocity $(\omega_{\ell b}^b)$ skew-symmetric matrix $\Omega_{\ell b}^b$ can be expressed in terms of its components as follows:

$$\Omega_{\ell b}^b = \Omega_{\ell i}^b + \Omega_{ib}^b = -\Omega_{i\ell}^b + \Omega_{ib}^b$$

$$\Omega_{\ell b}^b = \Omega_{ib}^b - \Omega_{i\ell}^b \qquad (5.33)$$

Hence, the rotation matrix R_b^{ℓ} is obtained by solving the following differential equation:

$$\dot{R}_b^{\ell} = R_b^{\ell}(\Omega_{ib}^b - \Omega_{i\ell}^b) \qquad (5.34)$$

Ω_{ib}^b is the skew-symmetric matrix of the angular velocity measurements ω_{ib}^b provided by the gyroscopes of the inertial system.

$$\omega_{ib}^b = \begin{pmatrix} \omega_x & \omega_y & \omega_z \end{pmatrix}^T \qquad (5.35)$$

where the skew symmetric representation is given as

$$\Omega_{ib}^b = \begin{pmatrix} 0 & -\omega_z & \omega_y \\ \omega_z & 0 & -\omega_x \\ -\omega_y & \omega_x & 0 \end{pmatrix} \qquad (5.36)$$

The gyroscopes measure the Earth's rotation rate, the change in orientation of the l-frame, and the angular velocities of the moving body. The angular velocities in Ω_{il}^b are subtracted from Ω_{ib}^b to remove the first two effects. Furthermore, the angular velocities Ω_{il}^b consist of two parts:

1. Ω_{ie}^b, which accounts for the Earth's rotation rate;
2. Ω_{el}^b, which accounts for the orientation change of the l-frame.

Therefore, Ω_{il}^b can be given in terms of its components as

$$\Omega_{il}^b = \Omega_{ie}^b + \Omega_{el}^b \qquad (5.37)$$

Ω_{ie}^b and Ω_{el}^b are skew symmetric angular velocity matrices corresponding to ω_{ie}^b and ω_{el}^b. Since the b-frame is not a reference frame, it is not possible to measure the angular velocities directly, and, as a result, transformation is required to compute the two vectors.

$$\omega_{ie}^b = R_\ell^b \omega_{ie}^\ell \qquad (5.38)$$

$$\omega_{el}^b = R_\ell^b \omega_{el}^\ell \qquad (5.39)$$

Finally, ω_{il}^b, which is used to construct the skew-symmetric representation Ω_{il}^b is written as:

$$\omega_{il}^b = \omega_{ie}^b + \omega_{el}^b = R_\ell^b(\omega_{ie}^\ell + \omega_{el}^\ell)$$

$$\omega_{il}^b = R_\ell^b \left[\begin{pmatrix} 0 \\ \omega^e \cos\varphi \\ \omega^e \sin\varphi \end{pmatrix} + \begin{pmatrix} \dfrac{-V^n}{M+h} \\ \dfrac{V^e}{N+h} \\ \dfrac{V^e \tan\varphi}{N+h} \end{pmatrix} \right] = R_\ell^b \begin{pmatrix} \dfrac{-V^n}{M+h} \\ \dfrac{V^e}{N+h} + \omega^e \cos\varphi \\ \dfrac{V^e \tan\varphi}{N+h} + \omega^e \sin\varphi \end{pmatrix} \qquad (5.40)$$

Equations 5.24, 5.29, and 5.34 define the navigation or mechanization equations in the *l*-frame and are usually written in the compact form as follows:

$$\begin{pmatrix} \dot{r}^\ell \\ \dot{V}^\ell \\ \dot{R}^\ell_b \end{pmatrix} = \begin{pmatrix} D^{-1}V^\ell \\ R^\ell_b f^b - (2\Omega^\ell_{ie} + \Omega^\ell_{e\ell})V^\ell + g^\ell \\ R^\ell_b(\Omega^b_{ib} - \Omega^b_{i\ell}) \end{pmatrix} \tag{5.41}$$

5.2.2 NED Realization

Mechanization equations for NED realization are slightly different as the definitions of the three axes (local *x*, *y*, and *z*-axes) have changed. The first change is that the *x*-axis of the ENU frame was east, which has now changed to north. Please note that the definitions of east and north have not changed. The position mechanization shows this change, and the second change is for the *z*-axis. The two *z*-axes have opposite directions. Only final mechanization equations for the NED frame are provided for clarity.

$$\dot{r}^\ell = \begin{pmatrix} \dot{\varphi} \\ \dot{\lambda} \\ \dot{h} \end{pmatrix} = \begin{pmatrix} \dfrac{1}{M+h} & 0 & 0 \\ 0 & \dfrac{1}{(N+h)\cos\varphi} & 0 \\ 0 & 0 & -1 \end{pmatrix} \begin{pmatrix} V^n \\ V^e \\ V^d \end{pmatrix} = D^{-1}V^\ell \tag{5.42}$$

The velocity dynamics equation is the same as (5.29). However, the angular velocities due to the Earth's and *l*-frame rotations differ as given in (5.43) to accommodate for the NED realization.

$$2\omega^l_{ie} + \omega^l_{el} = 2\begin{pmatrix} \omega^e\cos\varphi \\ 0 \\ \omega^e\sin\varphi \end{pmatrix} + \begin{pmatrix} \dfrac{V^e}{N+h} \\ \dfrac{-V^n}{M+h} \\ -\dfrac{V^e\tan\varphi}{N+h} \end{pmatrix} = \begin{pmatrix} \dfrac{V^e}{N+h} + 2\omega^e\cos\varphi \\ -\dfrac{V^n}{M+h} \\ -\dfrac{V^e\tan\varphi}{N+h} + 2\omega^e\sin\varphi \end{pmatrix} \tag{5.43}$$

The equation for attitude dynamics is also the same as (5.43). The rotation of the l-frame represented in the b-frame is changed due to the NED realization and is written as

$$
\omega_{il}^b = R_l^b \begin{pmatrix} \dfrac{V^e}{N+h} + \omega^e \cos\varphi \\ -\dfrac{V^n}{M+h} \\ -\dfrac{V^e \tan\varphi}{N+h} + \omega^e \sin\varphi \end{pmatrix} \tag{5.44}
$$

5.3 Solving Mechanization Equations

There are many methods that can be used to solve mechanization equations. Two methods are provided to demonstrate the implementation of the mechanization equations. The classical method is a simple method that is implemented by using (5.41). This method may be sufficient for low dynamic applications but is insufficient for any high dynamic application. The half-interval method, on the contrary, is a method that is robust, efficient, and is suited for most navigational scenarios.

5.3.1 Classical Method

This method is the most direct and straightforward implementation approach for the mechanization equations. Simplicity is guaranteed, but the accuracy of the navigation solution is not, even for land vehicle application. This method does not account for different types of motions that are related to the system and degrade the quality of results. In addition, this method cannot handle singularities. Details of the classical method are available in any standard navigation textbook [1], and here we will only provide Figure 5.4.

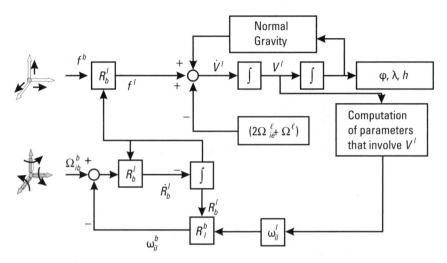

Figure 5.4 Flowchart for method 1.

5.3.2 Half-Interval Method

Half-interval method utilizes quaternions to increase the robustness and accuracies of the solution. A quaternion is a four-dimensional vector that can be used to represent a rotation matrix [6, 7]. Again this method is not discussed in this book since it has been covered in previous books and can be applied regardless of the inertial sensors used. The reader should seek more information through the references.

Chapters 3 and 4 illustrate the high error characteristics of these low-cost MEMS devices. Is there a way of reducing these errors? Can these errors be mitigaged by external aiding sources such as GPS or any other vehicle constraints? Chapter 6 answers these questions.

References

[1]   Titterton, D., and Weston, J.L., *Strapdown Inertial Navigation Technology, Second Edition*, Bodmin, U.K.: American Institute of Aeronautics and Astronomy (AIAA), 2004.

[2]   Schwarz, K.P., and Wei, M., "A Framework for Modelling the Gravity Vectory by Kinematic Measurements," *Bulletin Géodésique*, Vol. 64, 1990, pp. 331–346.

[3]   Farrell, J.A., and Barth, M., *The Global Positioning System & Inertial Navigation*, New York, NY: McGraw-Hill, 2001.

[4]   Godha, S., *Performance Evaluation of Low Cost MEMS-Based IMU Integrated with GPS for Land Vehicle Navigation Application*, M.Sc. Thesis, Department of Geomatics Engineering, University of Calgary, Canada, UCGE Report No. 20239, 2006.

[5]   Forsberg, R., "Gravity Field Terrain Effect Computations by FFT," *Bulletin Géodésique*, Vol. 59, 1985, pp. 342–360.

[6]   Dunn, F., and Parberry, I., *3D Math Primer for Graphics and Game Development*, Sudbury, MA: Wordware Publishing, Inc., 2002.

[7]   Kuipers, J.B., *Quaternions and Rotation Sequences*, Princeton, NJ: Princeton University Press, 1999.

[8]   Savage, P., *Strapdown Analytics,* Part 1 & 2, Maple Plain, Maple Plain, MN: Strapdown Associates, 2000.

[9]   Shin, E.-H., *Estimation Techniques for Low-Cost Inertial Navigation*, Ph.D. thesis, May 2005, Department of Geomatics Engineering, University of Calgary, Canada, UCGE Report 20219, 2005.

6

Aiding MEMS-Based INS

To mitigate high-error characteristics, MEMS inertial sensors require aiding sources for meaningful navigation results. If MEMS-based INS works in a standalone mode, positioning accuracy will deteriorate in a few minutes. It is also evident that a method is required that can propagate navigation states by incorporating the stochastic error characteristics of MEMS sensors. The problem of estimating the state of a stochastic (random variables evolving with time) dynamic system from noisy observation made on the system, is a vital research area of statistics, science, economy, and control engineering. In this case, an aiding source of superior accuracy is required to provide updates for the outputs of MEMS-based INS. This chapter briefly discusses the aiding sources and the modes of integration. In most applications, MEMS-based INS is integrated with GPS either through loosely or tightly coupled scheme. Both INS and GPS have complementary characteristics, and their integration provides reliable positioning accuracy. Extended Kalman filtering (EKF) is the most common INS/GPS integration technique used for this purpose. This chapter gives an overview of the integration between INS and GPS without going into too many details. Another source of updating MEMS-based INS is also described here briefly. A flowchart of the chapter is provided in Figure 6.1 as a quick reference to the user.

6.1 Introduction

Theoretically, INS and GPS can both estimate navigation parameters for a body in motion. However, both systems have their own problems. For example, the time-dependent position errors can drift quickly due to the integration of the acceleration and angular rate data for in-vehicle navigation. GPS-provided absolute and drift-free positions are only possible when the receiver has a direct line of sight to four or more satellites. The combination of the two systems can offer a number of advantages. The drift errors of the INS can be controlled by the GPS updates and for short GPS signal outages, the INS standalone navigation capabilities can be exploited for seamless navigation. Moreover, the combination of the two systems (i.e., INS and GPS) will provide redundant measurements and will result in improved reliability of the combined system.

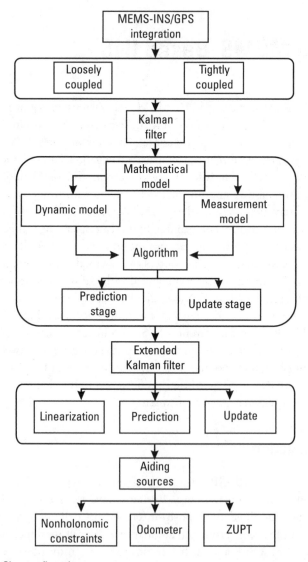

Figure 6.1   Chapter flowchart.

An EKF is used to optimally combine the redundant information in which the inertial state vector is regularly updated by GPS measurement. Two integration strategies can be implemented at the software level using the EKF approach. The remainder of this chapter is designed to provide the overview of the integration strategies, KF and EKF, for the most commonly encountered low-cost navigation scenario (i.e., GPS and MEMS-based IMU integrated navigation).

6.1.1 Loosely Coupled Mode of Integration

The most commonly implemented integration scheme is called the loosely coupled approach [1] in which the GPS derived positions and velocities, along with their accuracies from GPS KF, are used as updates for the navigation KF. The error states include both the navigation errors and sensor errors. To further improve the accuracy of the navigation solution, the error states are fed back to the mechanization [2] subroutine as shown in Figure 6.2.

There are certain advantages and disadvantages of using this integration scheme. For instance, one of the advantages is the smaller size of state vectors for both a GPS and INS KF as compared to the state vector in the tightly coupled integration. A disadvantage of using such a system is the extra process noise due to the presence of two KFs, which may decrease the signal-to-noise ratio. Consequently, the probability is that the integration filter trusts the predicted states more than the measurements increase, which is not desirable.

6.1.2 Tightly Coupled Mode of Integration

Tightly coupled integration [1, 2–4] is also known as the centralized KF approach. The major difference between loosely coupled (defined earlier) and tightly coupled is the number of KFs present in the two schemes. The tightly

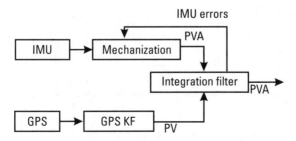

Figure 6.2 Loosely coupled integration. (PV refers to position and velocity, respectively.)

coupled integration uses one centralized KF [5, 6] that usually integrates the pseudorange (ρ) and Doppler (f_{dopp}) information from the GPS receiver and the position, velocity, and attitude (PVA) information from the mechanization of the inertial system. A tightly coupled integration with carrier phase observation [12] is mostly used for high-precision application, which is not the focus of this book.

The error states of the integration KF are composed of navigation errors, inertial sensor errors, and GPS receiver clock errors. The inertial sensor errors and GPS receiver clock errors are then fed back to compensate for these errors for the next epoch PVA estimation. Figure 6.3 shows the summary of the tightly coupled integration. The pseudorange (ρ) and Doppler (f_{dopp}) measurements from GPS, combined with the INS-derived pseudorange and Doppler for every satellite i, are used as the observations for the integration KF.

The loosely or tightly coupled integration scheme is realized by an EKF. It is the method of choice for blending inertial data with GPS updates for low-cost land vehicle navigation application due to its optimal weighting schemes.

6.2 Introduction to Kalman Filter

A KF is a clear choice for integration of inertial and GPS data under the assumptions of a Gaussian probability distribution, as outputs from both systems are contaminated with noise and uncertain dynamics [1, 4, 5]. The KF provides an efficient computational method to estimate the state of a linear stochastic process by minimizing the mean of the squared errors. The random

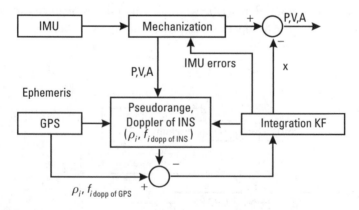

Figure 6.3 Tightly coupled integration scheme.

processes generally include noisy sensor outputs and uncertain dynamics. The uncertain dynamics include any form of disturbances other than the desired system states. For example, for the main application of this book (i.e., low-cost navigation for land vehicles) the uncertain dynamics may be due to road conditions or caused by unusual sensor outputs. An unusual sensor output for a MEMS sensor can be a change in the outputs due to temperature variations. This could be both significant and random depending on the type of low-cost sensor used.

The desired states of an integrated system generally consist of the navigation, clock, sensor parameters, and any other time-varying or correlated parameter that knowledge of may improve the overall navigation accuracy. The measurements may be the absolute positions, velocities, attitude, or pseudorange and Doppler information coming from the GPS, odometer, digital map, or an external attitude system.

The most desirable feature of the KF is its robustness. It comes from the fact that the KF estimates and propagates the uncertainty of the system states and measurements by using a gain equation. Kalman gain is the weighting matrix that combines the state and measurement uncertainties for the calculation of the updated system states as shown in (6.1).

$$\hat{x}_k^+ = \hat{x}_k^- + \bar{K}_k \left[z_k - H_k \hat{x}_k^- \right] \tag{6.1}$$

where $\hat{x}_k^+$ is the corrected system states at time epoch k, $\hat{x}_k^-$ is the estimated system states at time epoch k, $\bar{K}_k$ is the Kalman gain parameter, z_k is the measurements vector at time epoch k, H_k is the measurement matrix at time epoch k, and $H_k \hat{x}_k^-$ is the predicted measurement at time epoch k. However, knowledge of both mathematical models and the assumptions is required to understand (6.1). The details are provided in the following subsections.

There are two different mathematical models involved in the derivation of a KF:

1. The dynamic model that contains the time propagation information for the states; and

2. The measurement model that relates the measurements to the states.

6.2.1 Dynamic Model

The dynamic model [1] is the time-related information for a given state that shows how a state evolves with time. A first-order differential equation is used to show this type of time relation of a variable with respect to time for the state of a linear dynamic system.

$$\dot{x}(t) = F(t)x(t) + G(t)w(t) \tag{6.2}$$

where $F(t)$ is the dynamic matrix; $G(t)$ is the shaping matrix or noise distribution matrix, which represents how the sensor noise is distributed among the state vector parameters; and $w(t)$ is the sensor noise, which is considered as a white noise sequence with a mean of 0 and a covariance of Q.

The differential equation given above is used for a continuous time system. The measurements are available at certain fixed and finite frequency requiring discrete time implementation. The mathematics to convert the continuous time system to a discrete time system is beyond the scope of any navigation application and hence will not be discussed further.

For a discrete time system, the state (6.2) is given in terms of a transition matrix $\Phi_{k-1} = \Phi(t_k, t_{k-1})$. The transition matrix is then derived from the dynamic matrix F. A numerical approximation is possible if the dynamic matrix has constant values for a small time step Δt and therefore the impact of higher order terms is negligible.

$$\Phi_{k-1} = e^{F(t)\Delta t} = I + F(t)\Delta t + \left(F(t)\Delta t\right)^2 / 2! + \left(F(t)\Delta t\right)^3 / 3! + \dots \tag{6.3}$$

In the case of inertial measurements at data output rates of more than 20 Hz, the first two terms [in (6.3)] are sufficient to estimate the transition matrix. Using the newly computed transition matrix, the continuous time difference equation is given in the following form.

$$x_k = \Phi_{k-1}x_{k-1} + \int_{t_{k-1}}^{t_k} \Phi_{k-1}G(\tau)w(\tau)d\tau \tag{6.4}$$

The second term on the right side of this equation is an unknown and relates to the noise propagation in the discrete time system. Instead of solving this unknown, the whole term is considered as a random variable w_{k-1} with a Gaussian distribution, and for simplicity uses an additive noise [13]. If this Gaussian distribution assumption is not valid, the system suffers accuracy degradation depending on how far away the real distribution is compared to the assumed Gaussian distribution.

$$x_k = \Phi_{k-1}x_{k-1} + w_{k-1} \tag{6.5}$$

6.2.2 Measurement Model

The measurement model shows the relationship between the states and the available measurements using a measurement matrix $\mathbf{H}_k$. Most commonly, for vehicle navigation applications involving low-cost MEMS inertial sensors, the measurements z_k are the GPS updates at a much lower frequency. The basic linear measurement equation assuming additive noise for the KF has the following form:

$$z_k = \mathbf{H}_k x_k + v_k \tag{6.6}$$

where v_k is the white noise sequence for the measurements with covariance R_k.

In developing a measurement model, a reader must ask the following questions:

1. What kinds of updates or measurements are available?
2. Is there a direct relationship between the measurements and states?
3. What type of reference frame are the measurements given in as compared to the reference frame used for the navigation states?

The proper answers to these questions can satisfy the development of an appropriate model.

6.3 Kalman Filter Algorithm

A KF can be used whether or not a measurement update is present. This is a desirable property for integrated navigation because the data output rate for the inertial system is much faster than the most common updates (i.e., from GPS) [1]. The KF utilizes the update as soon as it becomes available. Consequently, the KF implementation is divided into two stages: the prediction stage and the update stage.

6.3.1 The Prediction Stage

This stage is implemented when there is no update available and the filter propagates the states and the states' accuracies from one epoch to the next using the dynamic matrix. If P_{k-1} is the a priori accuracy estimate for the states given by $\hat{x}_{k-1}^+$, the next P_k can be calculated using the transition matrix, the a priori accuracy matrix, and the covariance matrix for the involved system noise.

$$\hat{x}_k^- = \Phi_k \hat{x}_{k-1}^+ \tag{6.7}$$

$$P_k^- = \Phi_k P_{k-1}^+ \Phi_k^T + Q_{k-1} \tag{6.8}$$

6.3.2 The Update Stage

For navigation, the prediction stage relies on navigation states from the mechanization equations, which require integration. Any errors in the inertial sensor output, such as residual bias, accumulate during integration and result in a position drift error of hundreds of meters in short GPS outages when a low-cost MEMS IMU is used. Equation (6.1) is the update equation, which provides a robust blending of the prediction solution with the update measurements. This is due to the unique characteristics of Kalman gain, which weighs both the update and prediction.

$$\bar{K}_k = P_k^- H_k^T \left(H_k P_k^- H_k^T + R_k \right)^{-1} \tag{6.9}$$

This equation shows that a bigger uncertainty in the measurements makes the gain smaller while a smaller uncertainty in measurements makes the states to follow the measurements. Suppose that if the GPS accuracy is worse than the inertial-only navigation solution, the KF will prevent the navigation solution from following the GPS measurement update. This situation could occur if a high sensitivity GPS receiver is used in a downtown core, surrounded by the high-rises. There is a GPS position available, but the quality of the measurement might be poor. Similar to the prediction stage, the KF also calculates the accuracy of the newly updated states (6.10) by taking into account the Kalman gain.

$$P_k^+ = P_k^- - \bar{K}_k H_k P_k^- \tag{6.10}$$

6.4 Introduction to Extended Kalman Filter

The KF estimates the state of a discrete time controlled process governed by a linear stochastic difference equation. The linearity condition cannot be satisfied all the time and for all applications. The integration of inertial data with

GPS data using a KF is one example of when the system is nonlinear due to the involved mechanization equations. It is, however, not an isolated example and often the KF applications are nonlinear in nature.

For nonlinear cases, the system is linearized about a nominal trajectory during the design phase of the KF. For a general nonlinear case when the nominal trajectory is unavailable, the system can be linearized about the current state. In the case of inertial data integration, the current state is obtained by integrating the sensor output with respect to time using the mechanization process. A KF that involves linearization about the current state is referred to as an EKF.

6.4.1 Linearization

A navigation solution derived from the mechanization equations is a highly nonlinear problem, which requires linearization. To start, a simple dynamic nonlinear stochastic difference equation for the process with state x_k needs to be defined.

$$x_k = f(x_{k-1}) + w_{k-1} \tag{6.11}$$

The nonlinear difference equation given by the function f relates the previous epoch state x_{k-1} to the current epoch state x_k. The random variable w_{k-1} is the dynamic process noise with

$$E\left[w_k\right] = 0 \text{ and } E\left[w_k w_j^T\right] = Q_k \text{ for } k = j \tag{6.12}$$

Equation (6.11) is the simplest form of nonlinear difference equation where it is assumed that the noise is linear and only one measurement update is available at any given time. However, this may not be true for all cases. For example, there may be cases when the noise is nonlinear or there may be some redundant measurement updates present for the same navigation entity. Consider an example involving a single epoch with two heading measurements. The first heading is provided by a dual GPS system, while the second heading is provided by a magnetic compass. In this case, the accuracies for the two headings are different. One of the heading measurements is used as a driving function, u_{k-1} as shown in the nonlinear difference equation provided below, and the other heading is used as measurement update.

$$x_k = f(x_{k-1}, u_{k-1}, w_{k-1}) \tag{6.13}$$

Similarly, the measurements can also have a nonlinear relationship to the states. The simplest case is shown in (6.14) where noise is uncorrelated and Gaussian distributed.

$$z_k = h(x_k) + v_k \tag{6.14}$$

$$E\left[v_k\right] = 0, \qquad E\left[v_k v_j^T\right] = R_k \quad \text{for} \quad k = j \tag{6.15}$$

For a highly nonlinear model, the assumption of a linear noise component may not be realistic. In this case, even the noise needs to be linearized, and a better measurement model equation has the following form.

$$z_k = h(x_k, v_k) \tag{6.16}$$

For the prediction step, the noise cannot be estimated and is left out of the computation process on the assumption that it is a 0 mean with Gaussian distribution. Also, it is hardly the case that there are redundant measurements for the same state for very low-cost navigation systems and, therefore, it is safe to leave the driving function out of the dynamic equation linearization and from prediction. After making the above changes, the state and measurement vector approximates, $\tilde{x}_k$ and $\tilde{z}_k$ are given as

$$\tilde{x}_k = f\left(\hat{x}_{k-1}^-\right) \tag{6.17}$$

$$\tilde{z}_k = h\left(\tilde{x}_k\right) \tag{6.18}$$

For prediction and update equations, linearized models are needed. In EKF, the linearization is performed at the current state, which is the last available state vector. Taylor series expansion is used for linearization of the nonlinear difference equation as illustrated by (6.19).

$$x_k \approx f\left(\hat{x}_{k-1}^-, 0\right) + \partial f / \partial x \Big|_{\hat{x}_{k-1}^-} \left(x_{k-1} - \hat{x}_{k-1}^-\right) + (2!)^{-1} (\partial^2 f / \partial x^2)\Big|_{\hat{x}_{k-1}^-}$$
$$\left(x_{k-1} - \hat{x}_{k-1}^-\right)^2 + \ldots + \left\langle w_{k-1}\right\rangle \tag{6.19}$$

The quantities x_k and x_{k-1} are the *true* state vectors, which are unavailable. These values can only be estimated using the EKF, which introduces errors such as truncation errors. In short, only an estimate of the true state can be obtained using (6.20). The Jacobian matrix $\partial f / \partial x \Big|_{\hat{x}_{k-1}^-}$ that propagates a previous state vector to the current is the partial derivative of the nonlinear process function f with respect to the elements of the state vector x evaluated at the previous state estimate. For simplicity, this Jacobian matrix is referred to as $\mathbf{F}_k$. In the event that the process noise is also part of the nonlinear difference equation as shown in (6.13), the component of (6.19) given in $\langle\ \rangle$ are linearized by using the partial derivative of the function with respect to the noise vector w_{k-1} and evaluated at the estimated state $\hat{x}_{k-1}^-$ (i.e., $\langle W_k w_{k-1}\rangle$) where $W_k = \partial f /\partial x|_{\hat{x}_{k-1}^-}$. Therefore, the linearized form is written as

$$x_k \approx f\left(\hat{x}_{k-1}^-,0\right) + \partial f /\partial x\big|_{\hat{x}_{k-1}^-} \left(x_{k-1} - \hat{x}_{k-1}^-\right) + (2!)^{-1}\,\partial^2 f /\partial x^2\big|_{\hat{x}_{k-1}^-}$$
$$\left(x_{k-1} - \hat{x}_{k-1}^-\right)^2 + ... + \left\langle W_k w_{k-1}\right\rangle \tag{6.20}$$

Another linearization is necessary if the measurement equation is also nonlinear as discussed earlier. For the simple case given in (6.10), the Taylor series expansion has the following form.

$$z_k \approx \tilde{z}_k + \partial h /\partial x\big|_{\tilde{x}_k} \left(x_k - \tilde{x}_k\right) + (2!)^{-1}\,\partial^2 h /\partial x^2\big|_{\tilde{x}_k} \left(x_k - \tilde{x}_k\right)^2 + ... + v_k \tag{6.21}$$

The measurement vector z_k is the *true* measurement that can come from the GPS receiver or any other aiding sensors. Even a physical relationship can be used as the measurement [11]. The Jacobian matrix composed of a partial derivative of the measurement equation with respect to the state vector evaluated at the approximated current state is called H_k, i.e., $H_k = \partial h /\partial x|_{\tilde{x}_k}$.

As discussed earlier, for a highly nonlinear observation model, noise is a part of the nonlinear measurement function as given in (6.16). The linearization process should consider noise as the second parameter for linearization:

$$z_k \approx \tilde{z}_k + \partial h /\partial x\big|_{\tilde{x}_k} \left(x_k - \tilde{x}_k\right) + (2!)^{-1}\,\partial^2 h /\partial x^2\big|_{\tilde{x}_k} \left(x_k - \tilde{x}_k\right)^2 + ... + V_k v_k \tag{6.22}$$

where $V_k = \partial h /\partial v|_{\tilde{x}_k}$.

EKF implementation only considers the first-order approximation of the linearized dynamic process and measurement equations.

$$x_k \approx f\left(\hat{x}_{k-1}^-\right) + F_k \cdot \left(x_{k-1} - \hat{x}_{k-1}^-\right) + \left\langle W_k w_{k-1}\right\rangle \tag{6.23}$$

$$z_k = \tilde{z}_k + H_k \cdot \left(x_k - \tilde{x}_k\right) + \left\langle V_k v_k\right\rangle \tag{6.24}$$

where the $\cdot$ represents multiplication.

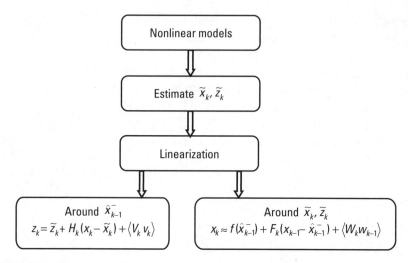

Figure 6.4   Linearization process for EKF.

The linearization process for the EKF discussion is summarized by Figure 6.4.

The information provided about KF and EKF is only a summary; for implementation the reader is encouraged to read [1, 7–9]. It should be noted that it is beyond the scope of this book to derive the dynamic error model of INS and the measurement model of GPS; however, they are shown in the Appendix.

6.4.2 EKF Limitations

The EKF assumes that all the noise components have a 0 mean and follow the Gaussian probability distribution function. However, it is not a valid assumption, especially when the noise is part of a nonlinear function. For example, GPS and INS integration using EKF consists of linearization of highly nonlinear dynamic equations. Unscented Kalman filters and particle filters are some of the other options that are available for nonlinear dynamics systems that do not require linearization.

In addition, the linearization process itself can contribute to significant linearization errors. EKF only takes the first-order linearization terms into account with the assumption that the higher-order terms are so small that they can be safely neglected without affecting the accuracy of the navigation states.

If this assumption fails, the navigation states are erroneous and the solution may become unreliable.

6.4.2.1 Tightly Coupled Verses Loosely Coupled Approach—An Example

In this example, a tightly coupled EKF was implemented. A small dataset that consists of four GPS signal degradation periods of approximately 50 to 70 seconds each is chosen for the processing. The first GPS degradation period consists of the time when the GPS receiver has a lock on only three satellites. The next degradation period has only two visible satellites, while the last two degradation periods have only 1 satellite signal due to trees and buildings. Figure 6.5 shows the results of the tightly coupled EKF.

As the tightly coupled EKF can use the raw GPS signals from any number of satellites in the integration, in theory, it should provide more accurate results than the loosely coupled integration technique. A loosely coupled EKF for this dataset generates four GPS signal outage periods instead of GPS signal degradation periods as there is no position information provided by the GPS receiver when the number of visible satellites is less than four. For comparison, the loosely coupled EKF results are provided in Figure 6.6.

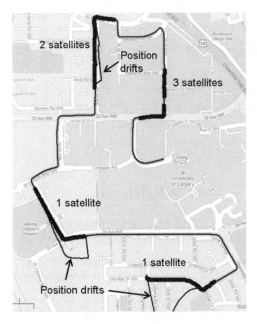

Figure 6.5 Tightly coupled integration for different GPS satellite availability.

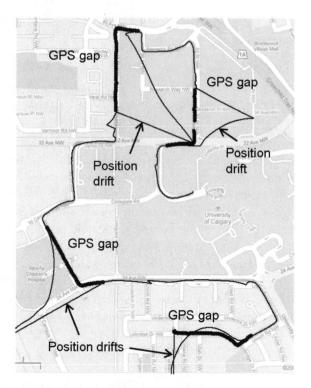

Figure 6.6   Loosely coupled integration with four GPS outages.

This chapter illustrates Kalman filters as the best solutions for multi-sensory integration. However, is KF an optimal filter under highly nonlinear conditions or with non-Gaussian noises? Chapters 7 and 8 answer this fundamental question.

References

[1]   Grewal, M.S., Weil, L.R., and Andrews, A.P., *Global Positioning Systems, Inertial Navigation and Integration*, New York, NY: John Wiley & Sons Inc., 2001.

[2]   Kim, J.W., Hwang, D.H., and Lee, S.J., "A Deeply Coupled GPS/INS Integrated Kalman Filter Design Using a Linearized Correlator Output," *PLANS 2006*, California, April 25–27, 2006, pp. 300–305.

[3]   Knight, D.T., "Rapid Development of Tightly Coupled GPS/INS Systems," *IEEE Aerospace and Electronic Systems Magazine*, Vol. 12, No. 2, February 1997, pp. 14–18.

[4]   Wendel, J., and Trommer, G.F., "Tightly Coupled GPS/INS Integration for Missile Applications," *Aerospace Science and Technology*, Vol. 8, No. 7, October 2004, pp. 627–634.

[5]   Hide, C., Moore, T., and Smith, M., "Adaptive Kalman Filtering for Low Cost GPS/IMU," *Journal of Navigation*, Vol. 56, No. 1, January 2003, pp. 143–152.

[6]   Godha, S., "Performance Evaluation of Low Cost MEMS-Based IMU Integrated With GPS for Land Vehicle Navigation Application," M.Sc. thesis, Department of Geomatics Engineering, University of Calgary, Canada, UCGE Report No. 20239, 2006.

[7]   Brown, R.G., *Introduction to Random Signal Analysis and Kalman Filtering*, New York: John Wiley & Sons, Inc., 1983.

[8]   Brown, R.G., and Hwang, P.Y.C., *Introduction to Random Signals and Applied Kalman Filtering, Second Edition*, New York: John Wiley & Sons, Inc., 1992.

[9]   Gelb, A., *Applied Optimal Estimation*, Cambridge, MA: MIT Press, 1974.

[10]   Featherstone, W.E., and Dentith, M.C., "A Geodetic Approach to Gravity Reduction for Geophysics," *Computers and Geosciences*, Vol. 23, No. 10, 1997, pp. 1063–1070.

[11]   Dissanayake, G., et al., "The Aiding of a Low Cost, Strapdown Inertial Unit Using Modelling Constraints in Land Vehicle Applications," *IEEE Trans. on Robotics and Automation*, Vol. 17, No. 5, 2001, pp. 731–747.

[12]   Wendel, J., and Trommer, G.F., "Tightly Coupled GPS/INS Integration for Missile Applications," *Aerospace Science and Technology*, Vol. 8, 2004, pp. 627–634.

[13]   Maybeck, P.S., *Stochastic Models, Estimation and Control, Volume 3*, Academic Press Inc., 1982.

7

Artificial Neural Networks

Despite the availability of several nonlinear filtering techniques, most of the present navigation sensor integration methods are still based on Kalman filtering estimation procedures mainly due to its simplicity. Although Kalman filtering represents one of the best solutions for multisensor integration, it still has some drawbacks in terms of stability, robustness, immunity to noise effects, and observability, especially when used with low-cost MEMS-based inertial sensors. Kalman filters perform adequately only under certain predefined dynamic models. If the Kalman filter is exposed to input data that does not fit the model, it will not result in reliable estimates. For low-cost INS/GPS integration modules, the nonlinear error terms are ignored when the linearized error model for Kalman filtering is developed, which contributes significantly towards large position errors, especially in the long term. Another problem related to the Kalman filter is the observability of the different states. The system is considered to be nonobservable if there are one or more state variables that are hidden from the view of the observer (i.e., the measurements). Consequently, if the unobserved process is not stable, the corresponding estimation errors will be similarly unstable. This chapter discusses the merits and limitations of different approaches developed for integrating INS and GPS using artificial intelligence.

7.1 Introduction

The artificial neural network (ANN) is adopted in the field of INS/GPS integration as either a total replacement for a Kalman filter or is combined with one for improving the overall positioning accuracy. ANN is a massively parallel distributed processor that allows modeling highly complex and nonlinear problems with a high level of stochastic that cannot be solved using conventional algorithms [1, 2].

ANNs are networks of many simple processors (units) operating in parallel, each possibly having a small amount of local memory. The units are connected by communication channels (connections), which usually carry numeric data (weight). The units operate only on their local data and on the inputs they receive via the connections. The restriction to local operations can often be relaxed during the learning process. An ANN should have specific training rules whereby the weights of connections are adjusted on the basis of learning data. In other words, an ANN learns from examples (of known input/output sequences) and exhibits some capability for generalization beyond the training data. An ANN normally has great potential for parallelism, since the computations of the components are largely independent of each other. The function of each element is determined by network structure, connection strengths, and the processing performed at computing elements or nodes.

The inspiration of ANNs came from the desire to produce artificial systems capable of performing sophisticated (or perhaps intelligent) computations that mimic the routine performance of the human brain. An ANN resembles the brain in two respects: first, that knowledge is acquired by the network through a learning process, and second, that the interneuron connection weights are used to store the knowledge. In a strict mathematical sense, ANNs do not provide a closed form solution for the problem but offer a complex, accurate solution based on a representative set of historical examples of the relationship [3]. An ANN is practically useful for classification and function approximation/mapping problems, which are tolerant of some imprecision and have a considerable amount of training data available. Almost any mapping between vector spaces can be approximated to arbitrary precision by feedforward ANNs (which are the type most often used in practical applications) if enough training data and computing resources are available.

Several ANN-based INS/GPS integration architectures utilizing different types of ANNs were suggested in the literature [11–21]. The different architectures vary on the types of inputs and outputs to the ANN module. Some of these architectures were examined using different types of ANNs including multilayer perceptron neural networks (MLPNN), radial basis function neural networks (RBFNN), and input delay neural networks (IDNN). Furthermore, the benefits of fuzzy reasoning were also explored and combined with the neural network offering adaptive neuro fuzzy inference system (ANFIS) method, which was also used for INS/GPS integration for different low-cost navigation applications. In general, there are two modes of operation for any of the above ANN types. The first is the update mode of operation where the empirical ANN model is established during a special training procedure that utilizes the INS and GPS data. As shown in Figure 7.1(a), the output of the ANN module is compared with a desired performance and the obtained error is fed to a certain training criterion, such that the ANN model parameters are tuned to minimize the error [usually the mean square error (MSE)]. After achieving the training goal, or once the desired response becomes unavailable (like in the case of GPS outage), an ANN operates in the prediction mode [shown in Figure 7.1(b)] where the ANN processes the input data and provides reliable estimates of the output. Figure 7.1 demonstrates the general concept of operation of the ANN for both the update and prediction modes, which are common for all INS/GPS integration architectures covered in this chapter. For each of these architectures, the type of ANN utilized and the training criterion used will be discussed in the following sections.

7.2 Types of ANNs

Among the various types of ANNs, the MLPNN and the RBFNN are widely used in several applications and have been frequently utilized for INS/GPS

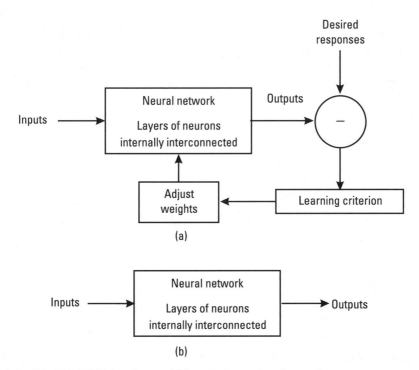

Figure 7.1 The ANN (a) update and (b) prediction modes of operation.

integration. Furthermore, because of its efficiency in dealing with uncertainty, imprecision, and vagueness in input data in dynamic environments, a fuzzy system was also employed by utilizing the ANFIS method for INS/GPS. The purpose of this section is to provide a brief background on MLPNN, RBFNN, and ANFIS as three types of ANNs widely used for INS/GPS integration purposes. More information can be found in [1–4].

7.2.1 Multilayer Perception Neural Network (MLPNN)

MLPNNs consist of several neuron layers with the different neurons of different layers connected to each other through certain weights. The structure of a neuron is shown in Figure 7.2. The inputs x_1, x_2, ... x_p are each transmitted through a connection that multiplies its strength by the scalar weight W. The weighted inputs are the only argument of the transfer function φ, which produces the output y. The threshold (or bias) θ may be viewed as simply being added to the weighted inputs. This sum is the argument of the transfer function φ, which is typically a step function or a sigmoid function. Note that W and θ are both adjustable scalar parameters of the neuron. The central idea of neural networks is that such parameters can be adjusted so that the network exhibits some desired or interesting behavior. Thus, the network can be trained to do a particular job by adjusting the weight or bias parameters, or perhaps the network itself will adjust these parameters to achieve some desired end.

Transfer functions φ for the neurons are needed to introduce nonlinearity into the network. Without this nonlinearity, neurons would perform in a

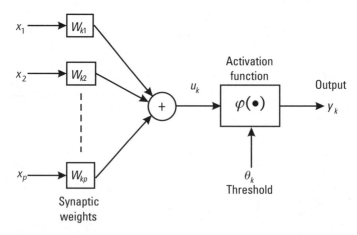

Figure 7.2   Single neuron model inside an MLPNN.

linear fashion and the MLPNN would not be able to map nonlinear input/ output relationships. For the output neuron(s), one should choose a transfer function suitable to the distribution of the target values. Bounded activation functions such as the logistic function are particularly useful when the target values have a bounded range. But if the target values have nonbounded ranges, it is preferable to use an unbounded activation function. If the target values are positive but have no upper bound, one can use an exponential output activation function. Many transfer functions have been introduced over the last few years by researchers specializing in ANNs. However, only three of these transfer functions are commonly used:

1. *The linear transfer function:* Neurons utilizing this transfer function are usually used as linear approximations in adaptive linear filters.

2. *The log-sigmoid transfer function:* This transfer function shown in Figure 7.3(a) takes the input, which may have an arbitrary value, and reduces the output into the range 0 to 1. This function is commonly used in backpropagation networks, in part because it is differentiable.

3. *The tan-sigmoid transfer function:* This transfer function is similar to the log-sigmoid except for the output, which can take values between ±1 as presented in Figure 7.3(b).

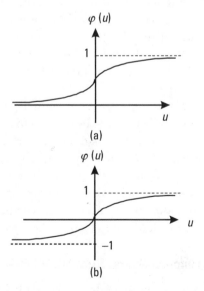

Figure 7.3 The log-sigmoid (a) and tan-sigmoid (b) transfer functions.

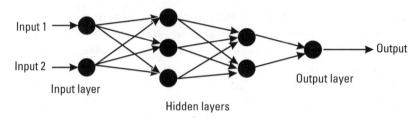

Figure 7.4 Example of an MLPNN.

Two or more of the neuron structures shown in Figure 7.2 may be combined to construct a neuron layer, and a particular network may contain one or more such layers. In general, the input and output layers of any network have a number of neurons equal to the number of the inputs and outputs of the system, respectively. The neuron layers between the input and the output layers are known as hidden layers. Figure 7.4 demonstrates an MLPNN of two hidden layers for a two inputs/one output problem. Examples and more implementation details are provided in Sections 7.3 and 7.4.

The number of neurons and hidden layers can be arbitrarily chosen and adjusted. It has been reported that one or two hidden layers with an arbitrarily large number of neurons may be enough to approximate any function [1]. MLPNNs often have one or more hidden layers of sigmoid neurons followed by an output layer of linear neurons. Multiple layers of neurons with nonlinear transfer functions allow the network to learn nonlinear and linear relationships between input and output data. The linear output layer lets the network produce values outside the range −1 to +1. To constrain the output of a network (such as between 0 and 1), a sigmoid transfer function (e.g., log-sigmoid) should be used at the output layer.

The learning criterion is a procedure for modifying the weights and biases of the network. This procedure may also be referred to as a training algorithm [1–5]. The learning rule is applied to train the network to perform some particular task. It is provided with a known input/output set of data. The known output data are considered as the target output of the network. As the inputs are applied to the network, the network outputs are compared to the targets. The learning rule is then used to adjust the weights and biases of the network in order to move the network outputs closer to the targets.

7.2.2 Radial Basis Function Neural Network (RBFNN)

The conventional neural network model of the MLPNN is based on units, which compute a nonlinear function of the scalar product of the input and weight vectors. An alternative architecture for ANNs is one in which the distance between the input vector and a certain prototype vector determines the activation of a hidden unit [4]. This architecture is known as a radial basis function neural network (RBFNN), which is composed of receptive units (neurons) that act as the operator providing the information about the class to which the input signal belongs. An RBFNN gives an approximation of any input/output relationship as a linear combination of the radial basis functions (RBF) [4, 6]. RBFs are a special class of functions with the characteristic feature that their response decreases (or increases) monotonically with distance from a central point. Although the architectural view of an RBFNN is very similar to that of an MLPNN, the hidden neurons possess basis functions to characterize the partitions of the input space. Each neuron in the hidden layer provides a degree of membership value for the input pattern with respect to the basis vector of the respective hidden unit itself. The output layer is comprised of linear neurons.

As shown in Figure 7.5, the structure of an RBFNN consists of an input layer, one hidden layer, and an output layer. The input layer connects the inputs to the network. The hidden layer applies a nonlinear transformation from the input space to the hidden space. The output layer applies a linear transformation from the hidden space to the output space.

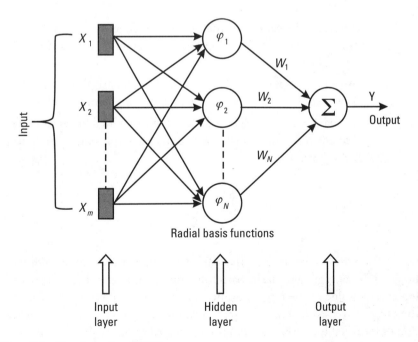

Figure 7.5 Structure of an RBFNN.

The radial basis functions φ_1, φ_2, ..., φ_N are known as hidden functions while $\{\varphi_i(x)\}_{i=1}^{N}$ is called the hidden space. The number of basis functions (N) is typically equal to or less than the number of data points available for the input data set. Among several radial basis functions, the most commonly used is the Gaussian, which in its one-dimensional representation takes the following form:

$$\varphi(x,\mu) = e^{-\frac{\|x-\mu\|^2}{2d^2}} \tag{7.1}$$

where μ is the center of the Gaussian function (mean value of x) and d is the distance (radius) from the center of $\varphi(x, \mu)$, which gives a measure of the spread of the Gaussian curve. The hidden units use the radial basis function. If a Gaussian function is used, the output of each hidden unit depends on the distance of the input x from the center μ. During the training procedure, the center μ and the spread d are the parameters to be determined. It can be deduced from the Gaussian radial function that a hidden unit is more sensitive to data points near the center. This sensitivity can be tuned (adjusted) by controlling the spread d. It can be observed that the larger the spread, the less the sensitivity of the radial basis function to the input data. RBFNNs require the following tasks during the training process [4]:

1. Allocating the centers of the radial basis functions as a part of the training procedure. Therefore, no constraints are imposed on these centers by the input vectors.

2. Instead of having a common spread d, each basis function is given its own spread that is determined during the training process.

3. The bias parameter W_0 is included in the weighted linear sum of the different basis functions provided at the output layer. The output $Y(x)$ is given as:

$$Y(x) = W_0 + \sum_{j=1}^{N} W_j \varphi_j(x) \tag{7.2}$$

where W_0 is the bias parameter. This parameter compensates for the difference between the average value of the basis function activations and the corresponding average value of the target output. W_j represents the vector of weights at the output layer as shown in Figure 7.5 where $(1 < j < N)$. Linear superposition of localized function (7.2) has been reported to be capable of developing a universal approximation for the output if the spreads of the Gaussian basis functions are treated as adjustable parameters [4, 6].

The unique architecture of RBFNNs makes their training procedure substantially faster than the methods used to train MLPNN. The interpretation given to the hidden units of an RBFNN leads to a two-stage training procedure. In the first stage, the parameters governing the basic functions (μ and d) are determined using fast unsupervised training methods that utilize only the input data. The second stage involves the determination of the output layer weight vector W_j. Since these weights are defined in a linear problem [see (7.2)], the second stage of training is also fast. It should be noted that the basic functions are kept fixed while the weights of the output layer are computed [4].

For INS/GPS integration, the RBFNN is trained (or updated) in real time during the availability of the GPS signal. Several learning algorithms have been proposed for training RBFNN. Among these different algorithms, the orthogonal least squares algorithm is the most popular. This algorithm considers that the number of radial basis functions in the hidden layer is known in advance. The centers of the radial basis functions $\{\mu_1, \mu_2, ..., \mu_N\}$ are chosen equal to a random subset of the input vector from the training set. The spread is determined by normalization—dividing the maximum distance between any two centers by the square root of the number of centers $\left(\dfrac{\text{Max}\left(\mu_j - \mu_k\right)}{\sqrt{N}} \right)$.

The optimal values of the weights at the output layer are found using the pseudoinverse method [4]. However, the choice of the number of radial basis functions (usually equal to the number of input data sets) before starting the training procedure may result in more radial basis functions than those really needed to develop an efficient RFBNN model. Using another approach [7], which is suitable for real-time implementation, consider only one radial basis function at the hidden layer, perform the training algorithm, and continuously repeat the algorithm while increasing the number of radial basis functions by one at each iteration. This iterative procedure stops after achieving a certain objective mean square estimation error. Such an approach would be computationally intensive since at each step it would be necessary to obtain a complete pseudoinverse solution to determine the weights at the output layer for each possible choice of the basis functions. In addition, the above technique may require a large amount of training data to achieve certain mean square estimation error. Extra details are provided in Sections 7.3 and 7.4.

In general, RBFNN is similar to MLPNN in providing techniques for approximating arbitrary nonlinear functional mappings between multidimensional spaces. However, the inherent structures of the two networks are very different. Some of these important differences that have made RBFNN more attractive than MLPNN for INS/GPS integration in real time are:

1. An MLPNN can have many layers of weights with a relatively complex pattern of connectivity. Several activation functions can be used within the same network. However, an RBFNN has a more simple architecture that consists only of two layers in which the first layer contains the parameters of the basis functions and the second layer forms linear combinations of these basis functions to generate the output. This unique architecture of the RBFNN has the advantage of a fast training procedure when compared to an MLPNN.

2. The parameters of the MLPNN (biases and weights) are determined simultaneously during the training procedure with supervised training techniques. However, an RBFNN is typically trained in two stages with the parameters of the basis functions being first determined by unsupervised learning techniques using the input data alone. The weights at the second layer are found by fast linear supervised methods.

3. The interference and crosscoupling between the different hidden units of an MLPNN may result in a highly nonlinear training process in addition to problems of local minima. This can lead to a very slow convergence even with advanced optimization strategies, which may not be appropriate for real-time implementation. By contrast, an RBFNN would not face this problem due to its localized basis functions, which are local with respect to the input space. Thus, for a given input vector, only a few hidden units will have significant activation.

7.2.3 Adaptive Neuro Fuzzy Inference System (ANFIS)

An ANFIS is a fuzzy logic algorithm that provides a method for mapping an input/output relationship based on available data tuples. The ANFIS architecture is based on the original Tagaki-Sugeno-Kang (TSK) fuzzy inference system [8–10]. The system acquires its adaptability by utilizing a hybrid learning method that combines backpropagation and least mean square optimization algorithms. When an adequate number of data tuples exist, learning can be achieved by tuning the membership functions using the gradient descent method to determine the premise parameters, along with applying the least mean square method to modify the consequent parameters so that the model output matches the system output with a minimum root mean square error. The ANFIS model was proven as an efficient mapping technique for many nonlinear systems. A basic example of an ANFIS structure used to provide the necessary data fusion for INS/GPS integration is presented in Figure 7.6 where a nonlinear model of INS error (for either position or velocity) can be established based on processing the time-varying INS position/velocity input. The model parameters can be obtained during the update mode of operation using special training algorithm while the GPS position and velocity outputs are available. In this example, two inputs [INS position (or velocity) and time] and one output [INS position (or velocity) error] are chosen. The first layer contains the membership functions. The number and shape of membership functions are predefined. However, the original spread and overlap of the membership functions is defined by the algorithm by using the fuzzy clustering algorithm. The membership function parameters are then tuned during the learning process. The fuzzy operator (Π) is operated at the second layer of the ANFIS and a normalized firing strength ($\overline{W}_i$) is computed in the third layer for each (i) consequent [8, 9].

The system output (i.e., INS error) is computed on the basis of the TSK fuzzy system as follows:

$$\text{if } x \text{ is } A_1 \text{ and } y \text{ is } B_1, \text{ then } f1 = p_1 x + q_1 y + r_1 \tag{7.3}$$

$$\text{if } x \text{ is } A_2 \text{ and } y \text{ is } B_2, \text{ then } f2 = p_2 x + q_2 y + r_2 \tag{7.4}$$

$$\text{INS\_error} = \Sigma w_i f_i \tag{7.5}$$

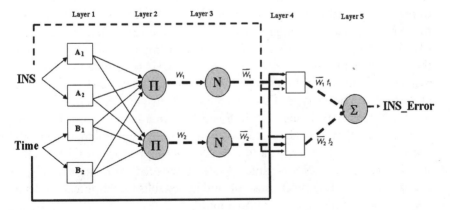

Figure 7.6 An example of an ANFIS structure for modeling INS errors.

where the linear parameters p_i, q_i, and r_i are the consequent parameters determined by the fuzzy system to relate the input fuzzy sets to the output fuzzy sets. The output in the fifth layer is the sum of the weighted outputs of all the fuzzy rules as in (7.5).

The major benefit of an ANFIS is its hybrid training methodology. It is hybrid since it utilizes least mean square optimization technique during forward path. Meanwhile, the descent gradient optimization technique is applied during backward computation propagation. For instance, during forward path, the fourth layer (rule consequents) parameters $\{p_i, q_i, r_i\}$ are determined. On the other hand, errors propagating during the backward propagation tune the membership function parameters (spread and center). This hybrid learning is very beneficial in terms of fast training and achievement of a low-error goal. This learning can be achieved by specifying a step size combined with increment and decrement rates that control the increase and decrease of the step size according to the root mean square error (RMSE) between ANFIS prediction and the system output. This allows the ANFIS to achieve a fast convergence towards the global minimum of the RMSE optimization function. Detailed descriptions about ANFIS parameters can be found in [8–10].

7.3 Whole Navigation States Architecture

The ANN is adopted in this architecture to mimic the vehicle dynamics and provide an INS/GPS integration module that processes the INS velocity and azimuth at its input and delivers the vehicle's position at the output [11–13]. The ANN is trained to perform a particular function by tuning the values of the weights (connections) between the neurons. An ANN has the capability to provide the necessary nonlinear processing of both INS and GPS despite the stochastic nature of their signals, which is the case when low-cost and MEMS-based inertial sensors are used. Another important advantage is that the ANN-based INS/GPS integration is a modeless technique that does not require any kind of prior knowledge. Thus, the proposed system is independent of the platform or the type of INS/GPS utilized. Two update architectures are described here, both integrating INS and GPS in a loosely coupled fashion where the ANN module is trained and the associated empirical model is established as long as the GPS data is available. During periods of GPS satellite signal blockage, the ANN module operates in the prediction mode relying on the output of the INS mechanization and the established empirical model to provide accurate positioning information.

7.3.1 Example of Position Update Architecture

The position update architecture (PUA) utilizes an MLPNN with a back-propagation training algorithm to integrate the data from INS and GPS and mimic the dynamical model of the moving vehicle carrying both systems [13]. After training the network, it can predict the vehicle's position during GPS signal blockage. As shown in Figure 7.7, the network inputs are the INS velocity (V_{INS}) and azimuth (ϕ_{INS}), which are obtained from the INS mechanization and produced at the INS data rate (typically 50 Hz). The output of the network is the difference in coordinates between two different consecutive epochs [$N(t)$ and $N(t-1)$ for the north position component and $E(t)$ and $E(t-1)$ for the east position component]. Position differences are used instead of the position component itself to simplify the training procedure. In fact, the magnitude of position components $N(t)$ and $E(t)$ are usually in the order of 10^5 meters, which may deteriorate the efficiency of the training algorithm to obtain the optimal values of the network weights. On the contrary, the magnitude of position differences are usually about 10^2 meters level and the relationship between the network inputs (INS velocity and azimuth) and the network outputs (position differences) is easier for neural networks to learn. Therefore, the utilization of position differences as network outputs reduces the complexity of the input/output function relationship, provides a more efficient MLPNN training, and reduces the time required for the training procedure. The INS velocity and azimuth information are used as inputs and the MLPNN outputs are compared to the GPS position differences. As long as the GPS signal is available, the training process continues to reduce the estimation error in order to obtain optimal values of the MLPNN weights. During GPS outages (i.e., signal blockage), the MLPNN parameters are used in prediction mode to provide estimates for the position components along the east and the north directions.

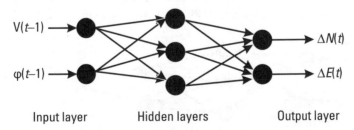

Figure 7.7 MLPNN module for PUA for INS/GPS integration.

7.3.2 Example of Position and Velocity Update Architecture

The position and velocity update architecture (PVUA) utilizes both the position and velocity information from GPS as the desired output. Thus, it contains two parallel processing architectures, PUA and velocity update architecture (VUA) [11–13]. Similar to PUA, the VUA utilizes a multilayer feedforward neural network (NN) with a backpropagation training algorithm. As shown in Figure 7.8, the VUA receives time information (t) and velocity outputs from INS [$V_{INS}(t)$] and generates estimated velocity outputs [$V_{VUA}(t)$], which are then compared to the GPS velocity information. As long as the GPS signal is available, the training process is continuously improving the estimation error in order to obtain the optimal values of the VUA network weights. During GPS outages, the MLPNN parameters are used in prediction mode to compensate for INS velocity errors and to provide accurate estimates for the velocity components.

It should be noted that the above two architectures (PUA and PVUA) are based on establishing an empirical model for the INS position and velocity while the GPS signal is available. Neither provides a prediction of the position and velocity errors like the case of Kalman filtering. The training procedure can be established using a special windowing technique like the one described in [4] to enable near real-time operation of the update (training) procedure.

Both PUA and PVUA approaches showed superior performance over KF during long periods of GPS outages with at least 70% improvement in the case of PUA and at least 80% improvement in the case of PVUA. However, they were proposed originally for high-grade inertial sensors (< 0.3°/rt hr) and have not been examined for very low-cost inertial sensors.

7.4 Navigation Error States Architecture

INS/GPS integration algorithms based on MLPNNs have been suggested and applied to different high-grade INS. It has been shown that a PVUA utiliz-

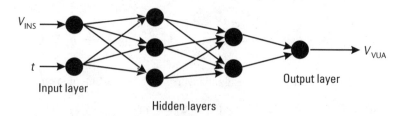

Figure 7.8 MLPNN module for VUA for INS/GPS integration.

ing two MLPNNs can process the INS azimuth and velocity and provide the position components along both the east and north directions. The parameters of the MLPNNs are adapted using GPS position and velocity updates. However, the PVUA system did not provide an update scheme for the vertical position components (the vehicle altitude). In addition, the ANN model of PVUA deals with the INS position components instead of their errors. Thus, no information about the accuracy of this system and the statistical properties of the estimates can be delivered during the navigation mission. Furthermore, besides the fact that it is almost impossible to conduct sensitivity analysis for the effect of changing the internal structure (number of hidden layers and number of neurons in each layer) of the MLPNNs on the system performance, the real-time implementation and the accuracy of the system during this mode of operations cannot be addressed.

In an attempt to design a modeless module that operates similarly to a KF but with no need for dynamic or stochastic models for INS, the $P - \delta P$ approach was proposed [14–17]. An MLP network was initially suggested to be used for each position component that processes the INS position (P) at the input and provides the corresponding INS position error (δP) at the output [14]. The suggested method updated the three $P - \delta P$ networks (for the three position components) by GPS position updates. Although information about the accuracy achieved during the navigation mission became available, the internal structure of each of the MLP networks has to change until the best performance is achieved, which leads to a long design time. In addition, the issue of the real-time implementation continues to be a challenge. The $P - \delta P$ architecture can be improved by using RBFNNs instead of MLPNNs [15, 16] since RBFNNs can be utilized without identifying the number of neurons in their hidden layer. In addition, hidden neurons are dynamically generated during the training procedure to achieve the desired performance. The original major drawback of this RBFNN-based module was that the training was still performed using all INS and GPS data available prior to the GPS outage, which was determined to be impractical and almost impossible to implement in real time because of long training times. In addition, the real-time implementation and the factors that affect the performance of the system during this mode of operation have not been addressed in the online RBF-based $P - \delta P$ module. The real-time realization issue can be addressed using a special windowing technique that enables near real-time operation of the RBFNN modules [16, 17]. The ANFIS can be also used with the $P - \delta P$ module for INS/GPS integration [18]. However, the optimization of the ANFIS parameters during the operation of the system may pose more computation load for real-time implementation. The ANFIS-based $P - \delta P$ architecture is improved when utilizing both GPS position and velocity updates with real-time crossvalidation during the update mode of operation to ensure the generalization of the model.

7.4.1 Architecture for INS/GPS Integration

The INS/GPS integration based on the $P - \delta P$ architecture is based on the estimation of INS position error (δP_{INS}) by processing the INS position (P_{INS}). The proposed module can utilize either an ANFIS or RBFNN to provide an optimal temporal estimation of the INS errors (δP_{INS}). An MLPNN is usually avoided. The proposed system architecture is comprised of three modes of operation: *initialization*, *update* [Figure 7.9(a)], and *prediction* [Figure 7.9(b)]. The initialization mode and the update mode are utilized to initialize the first learning rule base and to limit INS error growth, respectively, during GPS signals availability. The prediction mode is used to correct the INS position when the GPS signal is lost. Thus, the $P - \delta P$ module is trained during the availability of the GPS signal to recognize patterns of the position error embedded in the input position components. In the case of satellite signal blockage, the $P - \delta P$ module mimics the latest vehicle dynamics and delivers a prediction of the vehicle position error.

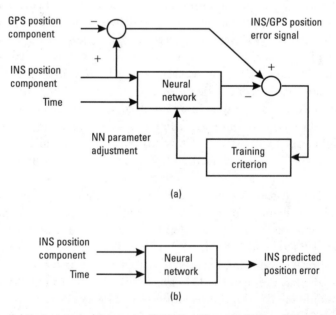

Figure 7.9   (a) Update mode of operation (GPS signal is available), and (b) prediction mode of operation (during GPS outages) for the $P - \delta P$ architecture.

The INS position (P_{INS}) and time (T) are the inputs to the module while the error in the INS position (δP_{INS}) is the module output. The estimated INS position error provided by the $P - \delta P$ module is then compared to the error between the INS original position and the corresponding GPS position ($\delta P_{INS|GPS}$). The number and shape of the radial basis functions when using RBFNN, or the membership functions when using ANFIS, shall be predefined. (Gaussian shape membership functions are usually used; however, other shapes may also be possible.) The original mean and spread of the membership functions are computed as part of the training criterion. For an ANFIS, the membership values are evaluated at the first layer and the fuzzy t-norm operator (Π) is implemented at the second layer. A normalized firing strength is computed at the third layer. Details of the training procedure can be further explored in [3, 5, 9].

The difference between δP_{INS} and $\delta P_{INS|GPS}$ is the training error $\Delta(\delta P)$ of the RBFNN or ANFIS module. For an RBFNN, in order to minimize this error, the training of the RBFNN module continues and the RBFNN parameters are continuously updated according to the least square criterion until reaching a certain minimal mean squared estimation error (MSE). To minimize the training error, the ANFIS module updates the learning rule base (membership function parameters and consequent parameters) using a hybrid learning approach combining the least square and backpropagation techniques until reaching a certain minimal RMSE.

When the satellite signal is blocked (during GPS outages), the system is switched to the prediction mode where the $P - \delta P$ module is used to predict the INS position error using the latest learning rule base obtained before losing the satellite signals. The error is then removed from the corresponding INS position component to obtain the corrected INS position (P_{INSC}). To obtain a three-dimensional positioning solution, three $P - \delta P$ modules are developed to provide a complete navigation solution in the three axes for a moving vehicle represented by the three position components.

Several other sources of errors should also be highlighted here that contribute to the overall position error of the vehicle in different ways. As discussed in Chapter 3, these sources include the inertial sensor bias drift and scale factor instability, the initial misalignment error during INS initialization, and the error in the heading angle of the vehicle (the azimuth error). The azimuth error, for example, can have a significant contribution on the position error along both the east and north position components, especially at high speeds since it is modulated by the vehicle velocity. Although all these error sources are not explicitly represented in the architecture shown in Figure 7.9, it is evident that the $P - \delta P$ module incorporates the effects of such errors in two ways. First, it establishes the learning rule base to pattern vehicular navigation performance using exemplar navigation scenarios that have been affected by the same set of error sources. Second, it considers the input parameters as interval data with membership values rather than crisp values, which provides room for uncertainty in the input parameters due to these sources of errors.

7.4.2 System Implementation

In order to utilize the $P - \delta P$ module in a temporal INS/GPS integration, a sliding window with a certain window size (W) is considered. For each of the three RBFNN/ANFIS modules, the number of samples (equal to W) of the INS position component P_{INS} and the corresponding GPS position P_{GPS} are acquired from both systems. The INS position is considered as the input to the $P - \delta P$ module and the error between P_{INS} and P_{GPS} is considered to be the corresponding desired response ($\delta P_{INS|GPS}$). The update procedure of the $P - \delta P$ learning rule base starts after collecting the Wth sample of both INS and GPS position components. Before considering the next INS and GPS samples, the ANFIS module is trained to mimic the position errors until reaching a certain minimum RMSE or after completing a certain number of training epochs. To guarantee timely operation of the system, the update procedure is terminated at the end of this number of training epochs even if the desired RMSE is not achieved.

While the GPS signal is available, the data window continues to slide, collecting new samples from INS and GPS position components. The erroneous INS signal has to be corrected before being applied to the $P - \delta P$ module. Therefore, assuming $\delta P_{INS}(i + 1) = \delta P_{INS}(i)$, the corrected INS position component can be obtained as

$$P_{INSC}(i + 1) = P_{INS}(i + 1) - \delta P_{INS}(i) \qquad (7.6)$$

where $\delta P_{INS}(i)$ is the INS position error provided by the $P - \delta P$ module at time instance i.

Since the GPS signal is available during the update mode, both P_{INSC} and the position error between P_{INSC} and P_{GPS} are used to train the ANFIS to mimic the dynamics present within the last data window. This results in a new ANFIS rule base, which is used to provide an estimate for the INS position error at time $i + 1$ [$\delta P_{INS}(i + 1)$]. Therefore, the INS position is continuously updated and the INS position errors (estimated by the ANFIS module) are removed, thus keeping accurate INS position components available in case of any GPS outages.

It should be noted that the proper choice of the window size is essential to guarantee the desired accuracy while ensuring system robustness in real time. The complexity of choosing the window size is related to its dependence on the level of vehicle dynamics and the length of the outage periods and its significance on the update procedure. Therefore, there is a trade-off in choosing a small or large window size. Large window sizes allow mimicking of significant details of the latest vehicle dynamics; thus, the module becomes reliable during GPS outages. However, large window sizes may complicate the update procedure and result in a long processing time. On the other hand, a fast and robust update procedure can be achieved using small window sizes due to the reduced level of nonstationary INS and GPS data. Moreover, using relatively small window sizes prevents the consideration of inaccurate position information provided by the ANFIS module during the first outage for the prediction of the INS position components. However, relatively small window sizes may cause the system to lose reliability in the case of relatively long GPS outages. As a trade-off to the problem, optimal window sizes can be determined using means of multiobjective optimization (considering reliability and updating time requirements) or by extracting a set of heuristic learning rules based on system observation.

7.4.3 The Combined $P - \delta P$ and $V - \delta V$ Architecture for INS/GPS Integration

In the $P - \delta P$ architecture from Section 7.4.1, the ANN module is trained to mimic the latest vehicle dynamics so that, in the case of GPS outages, the system relies on INS and the recently updated ANN module to provide the vehicle position. Several neural networks and neurofuzzy techniques were implemented in a decentralized fashion and provided acceptable accuracy for short GPS outages. In some cases, poor positioning accuracy during relatively long GPS outages may be obtained. In order to prevail over this limitation, $P - \delta P$ architecture can be further optimized by utilizing both GPS position and velocity updates. The ANN module (either ANFIS or RBFNN) for INS/GPS integration is designed to work in real time to fuse the INS and the GPS position and velocity data, to estimate the INS position and velocity errors, and to enhance the INS positioning accuracy [19]. Along each of the three axes of the navigation frame, two ANN modules are utilized. One of the modules is for the position and the other is for the velocity. For each of these modules, two inputs (INS position/velocity and time) and one output (INS position/velocity error) are used.

The ANN velocity module processes the INS velocity (V_{INS}) and time (t) as inputs and provides the corresponding error in the INS velocity (δV_{INS}) at the output. The estimated INS velocity error is then compared to the error between the INS original velocity and the corresponding GPS velocity ($\delta V_{INS|GPS}$). The difference between δV_{INS} and $\delta V_{INS|GPS}$ is the estimation error [$\Delta(\delta V)$] of the ANN velocity module. In order to minimize this error, the ANN velocity module is trained to adjust the ANN parameters that are continuously updated according to the least square criterion until reaching a certain minimal RMSE error. However, when the GPS signal is blocked, the system is switched to the prediction mode where the ANN velocity module is used to predict the INS velocity errors (δV_{INS}) using the latest ANN parameters obtained before losing the satellite signals. The error is then removed from the corresponding INS velocity component to obtain the modified INS velocity ($\hat{V}$). The position update procedure along the three directions follows the velocity update procedure to obtain the three position components based on integrating the corrected INS velocities. The modified INS position (P^{m}_{INS}) and time (t) are the inputs to the ANFIS position module while the error in the INS position (δP_{INS}) is the module output. The INS position error estimated by the ANN module (δP_{INS}) is then compared to the error between the INS-modified position and the corresponding GPS position ($\delta P_{INS|GPS}$). The difference between δP_{INS} and $\delta P_{INS|GPS}$ is the estimation error [$\Delta(\delta P)$] of the ANFIS position module. To minimize this error, the ANN module is trained to update the rule-base parameters that describe the fuzzy system premises and consequences, such that the RMSE is minimized. However, when the satellite signal is blocked (during GPS outages), the system is switched to the prediction mode where the ANFIS position module is used to predict the INS position error using the latest ANN parameters obtained before losing the satellite signals. The error is then removed from the corresponding INS-modified position to obtain the corrected INS position ($\hat{P}$). A block diagram of the above ANN-based approach for INS/GPS integration is shown in Figure 7.10.

Although the suitability of the previously mentioned techniques for several high grades of INSs is a proven fact, several challenges and inadequacies have been observed when they are applied to low-cost INS, especially for relatively long GPS outages in the long-term part of the trajectory [20, 21]. The high noise level and large bias drifts associated with very low-cost MEMS inertial sensors result in a large drift in the position and velocity components when MEMS-based INS operates in the standalone mode. Augmentation with Kalman filtering usually makes the ANN-based architectures in Sections 7.3–7.4 more suitable for such systems.

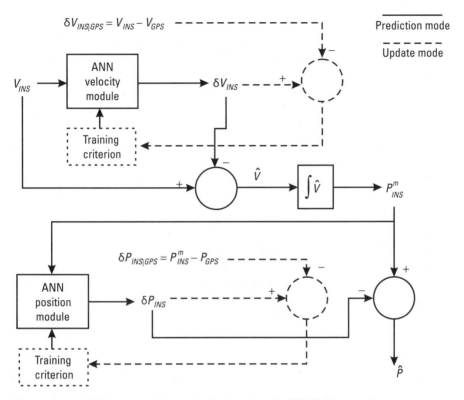

Figure 7.10 ANFIS-based position and velocity update for INS/GPS integration.

7.4.4 ANN/KF Augmented Module for INS/GPS Integration

For very low-cost MEMS-based INS/GPS integration, the $P - \delta P$ and $V - \delta V$ architectures can be further enhanced if augmented with Kalman filtering. This provides robust nonlinear modeling of INS errors and accurate positioning information in cases of long GPS outages. The major limitation of the above ANN-based modules is that they cannot be used standalone with very low-cost MEMS-based sensors without the aid of Kalman filtering as the output of the low-cost sensors degrades quickly over time and must be corrected at regular intervals. The growing positional error affects the performance of the ANN module, since after only a short period of time the mechanized IMU output is no longer reliable.

The INS positions/velocities along each direction are corrected with the errors computed at the prediction stage of the KF. Each of these corrected positions/velocities is used as an input to an ANN module (either RBFNN or ANFIS). The output of each module is the corresponding residual INS position/velocity error. During the update procedure, the three ANN modules are updated (trained) using the GPS position/velocity. This position/velocity update scheme creates an empirical model for the stochastic, nonstationary, and nonlinear parts of the INS errors that cannot be removed by a KF. These errors cannot be efficiently estimated at the KF prediction stage because it relies totally on the INS error model. In urban canyons or any denied GPS environments, the three ANN modules operate in the prediction mode. The three INS position/velocity components are processed by removing the corresponding errors computed by the KF prediction stage and the ANN modules. The output of the prediction stage of the KF is called the *predicted INS error* and it is the input to the AI module. The INS position/velocity without the predicted INS error is called the *predicted INS position/velocity*. The output of the AI module is called the *corrected INS error* and is compared to the *target position/velocity error*, which is the difference between the predicted INS position/velocity and the GPS position/velocity. Figure 7.11 demonstrates the block

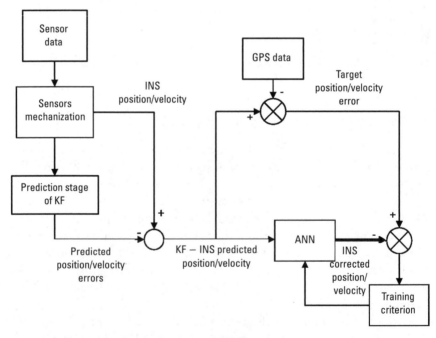

Figure 7.11   Flowchart of the augmented ANN/KF architecture.

diagram of the augmented ANN/KF solution for both position and velocity updates from GPS. In this figure, the ANN module is trained to model the residual nonlinear parts of the INS errors. For very low-cost MEMS-based inertial systems, those nonlinear error terms neglected for the dynamic error model used by KF may significantly deteriorate the performance. When the MEMS-based INS operates in the standalone mode during the GPS satellite signal blockage, the large values of stochastic bias drifts and scale factor instabilities may make these nonlinear error terms significant, thus deteriorating the overall system accuracy since they are not accounted for by KF. In the architecture shown in Figure 7.11, the augmentation of the ANN with the KF provides an empirical model for the nonlinear error terms and enhances the overall system performance. Should a GPS outage occur, the KF operates in the prediction mode predicting the linear part of the error, which is used to correct the output of the mechanization algorithm before being fed to the ANN module to remove the residual nonlinear error parts and provides the overall corrected position/velocity.

This chapter illustrates artificial neural networks as a multisensor data integration tool. However, is there any other kind of approximation or sampling filter available for integration? This question is answered in Chapter 8.

References

[1] Haykin, S., *Neural Networks—A Comprehensive Foundation*, Piscataway, NJ: IEEE Press, 1994.

[2] Ham, F.M., and Kostanic I., *Principles of Neurocomputing for Science and Engineering*, New York: McGraw-Hill, 2001.

[3] Tsoukalas, L.H., and Uhrig, R.E., *Fuzzy and Neural Approaches in Engineering*, New York: Wiley, 1997.

[4] Bishop, C.M., *Neural Networks for Pattern Recognition*, U.K.: Oxford University Press, 1995.

[5] Ross, T.J., *Fuzzy Logic with Engineering Applications*, West Sussex, U.K.: Wiley, 2004.

[6] Hui, P., et al., "A Parameter Optimization Method for Radial Basis Function Type Models," *IEEE Transactions on Neural Networks*, Vol. 14, No. 2, 2003, pp. 432–438.

[7] Karayiannis, N.B., and Randolph-Gips, M.M., "On the Construction and Training of Reformulated Radial Basis Function Neural Networks," *IEEE Transactions on Neural Networks*, Vol. 14, No. 4, 2003, pp. 835–846.

[8] Jang, J.S.R., Sun, C.T., and Mizutani, E., *Neuro-Fuzzy and Soft Computing, a Computational Approach to Learning and Machine Intelligence*, Englewood Cliffs, N.J.: Prentice Hall, 1997.

[9]   Jang, J.S.R., "ANFIS: Adaptive Network-Based Fuzzy Inference Systems," *IEEE Transactions on Systems, Man and Cybernetics*, Vol. 23, No. 3, 1993, pp. 665–685.

[10] Loukas, Y.L., "Adaptive Neuro-Fuzzy Inference System: An Instant and Architecture-Free Predictor for Improved QSAR Studies," *J. Med. Chem.*, Vol. 44, No. 17, 2001, pp. 2772–2783.

[11] Chiang K.-W., Noureldin, A., and El-Sheimy, N., "Multi-Sensors Integration Using Neuron Computing for Land Vehicle Navigation," *GPS Solutions*, Vol. 6, No. 4, 2003, pp. 209–218.

[12] Chiang K.-W., Noureldin, A., and El-Sheimy, N., "A New Weights Updating Method for Neural Networks Based INS/GPS Integration Architectures," *Measurement Science and Technology, IoP*, Vol. 15, No. 10, 2004, pp. 2053–2061.

[13] Chiang, K.-W., Noureldin, A., and El-Sheimy, N., "The Utilization of Artificial Neural Networks for Multi-Sensor System Integration in Navigation and Positioning Instruments," *IEEE Transactions on Instrumentation and Measurement*, Vol. 55, No. 5, 2006, pp. 1606–1615.

[14] Noureldin, A., Osman, A., and El-Sheimy, N., "A Neuro-Wavelet Method for Multi-Sensor System Integration for Vehicular Navigation," *Measurement Science and Technology, IoP*, Vol. 15, No. 2, 2004, pp. 404–412.

[15] Sharaf, R., et al., "Online INS/GPS Integration with a Radial Basis Function Neural Network," *IEEE Aerospace and Electronic Systems*, Vol. 20, No. 3, 2005, pp. 8–14.

[16] Sharaf, R., and Noureldin, A., "Sensor Integration for Satellite Based Vehicular Navigation Using Neural Networks," *IEEE Transaction on Neural Network*, Vol. 18, No. 2, 2007, pp. 589–594.

[17] Semeniuk, L., and Noureldin, A., "Bridging GPS Outages Using Neural Network Estimates of INS Position and Velocity Errors," *Measurement Science and Technology, IoP*, Vol. 17, No. 9, 2006, pp. 2782–2798.

[18] Sharaf, R., et al., "Merits and Limitations of Using Adaptive Neuro-Fuzzy Inference System for Real-Time INS/GPS Integration in Vehicular Navigation," *Soft Computing*, Vol. 11, No. 6, 2007, pp. 588–598.

[19] Noureldin, A., El-Shafie, A., and El-Sheimy, N., "Adaptive Neuro-Fuzzy Module for INS/GPS Integration Utilizing Position and Velocity Updates with Real-Time Cross Validation," *IET Radar, Sonar and Navigation*, Vol. 1, No. 5, 2007, pp. 388–396.

[20] Perreault, J., et al., "RISS/GPS Integration Utilizing an Augmented KF/NN Module," *European Journal of Navigation*, Vol. 6, No. 3, 2008, pp. 15–21.

[21] Noureldin, A., et al., "Performance Enhancement of MEMS Based INS/GPS Integration for Low Cost Navigation Applications," *IEEE Transactions on Vehicular Technology*, Vol. 58, No. 3, 2009, pp. 1077–1096.

8

Particle Filters

Particle filters (PF) are capable of handling highly nonlinear models with any kind of noise distribution. These filters are based on the sequential Monte Carlo principle that represents the required noise or data distribution by a cluster of samples called particles. Before going any further, a flowchart of the chapter is provided in Figure 8.1.

8.1 Introduction

The objective of filtering is to estimate the state vector x_k, conditioned upon all the measurements up to the current time $y_{1:k}$ [1–3]. To best describe the random nature of the involved state variables, probability density functions (PDFs) are used. The PDFs are constantly changing shape when receiving indirect measurements y_k. The PDFs will evolve with time and have to be updated recursively upon receiving new information. The estimation problem involves calculation of the posterior density function $p(x_{0:k} \mid y_{1:k})$ based on Bayes' theorem [as given in (8.1)] and all currently available data.

$$p\left(x_{0:k} \mid y_{1:k}\right) = \frac{p\left(y_{1:k} \mid x_{0:k}\right) p(x_{0:k})}{p\left(y_{1:k}\right)} \tag{8.1}$$

where $p(x_{0:k})$ is the prior density function that gives information about the state vector $(x_{0:k})$ without any knowledge of the measurements; $p(y_{1:k} \mid x_{0:k})$ is the likelihood functions of the measurement vector $(y_{1:k})$ given the states $x_{0:k}$; $p(x_{0:k} \mid y_{1:k})$ tell us what is known about $x_{0:k}$ when measurements are available, and hence is called the posterior density function of $x_{0:k}$ given $y_{1:k}$.

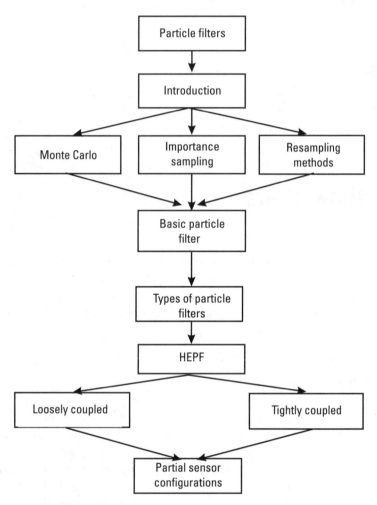

Figure 8.1   Flowchart for Chapter 8.

The denominator $p(y_{1:k})$ is the normalization constant and for a discrete distribution is equal to $\Sigma p(y_{1:k} \mid x_{0:k})p(x_{0:k})$. According to the first-order Markov assumption, the state vector is conditionally independent of all earlier states given the immediately previous state; the prior density function is factorized as $p(x_{0:k}) = p(x_0)\prod_{i=1}^{k} p(x_i \mid x_{i-1})$. If measurements $(y_{1:k})$ are independent given states $(x_{0:k})$ (i.e., y_k does not depend on previous $y_{1:k-1}$), we get the simplified likelihood function as $p\left(y_{1:k} \mid x_{0:k}\right) = \prod_{i=1}^{k} p\left(y_i \mid x_{0:k}\right)$. Furthermore, since measurements are conditionally dependent only on the current state, the likelihood function becomes $p(y_{1:k} \mid x_{0:k}) = \prod_{i=1}^{k} p(y_i \mid x_i)$. The simplified posterior density function, upon incorporating these changes, is represented by (8.2).

$$p(x_{0:k} \mid y_{1:k}) = \frac{p(x_0)\prod_{i=1}^{k} p(y_i \mid x_i)p(x_i \mid x_{i-1})}{p(y_{1:k})} \tag{8.2}$$

Equation (8.2) requires all states to be stored. To sequentially estimate $p(x_{0:k} \mid y_{1:k})$, propagation needs to occur without modifying the past simulated states [i.e., a modified (8.2) is needed as provided by (8.3)].

$$p(x_k \mid y_{1:k}) = p(x_{k-1} \mid y_{1:k-1})\frac{p(y_k \mid x_k)p(x_k \mid x_{k-1})}{p(y_k \mid y_{1:k-1})} \tag{8.3}$$

where $p(x_{k-1} \mid y_{1:k-1})$ is the previous posterior density function, $p(y_k \mid x_k)$ is the likelihood function, $p(x_k \mid x_{k-1})$ is the state transition prior probability density function, and $p(y_k \mid y_{1:k-1})$ is the normalization constant [1]. The marginal density function or the prior distribution $p(x_k \mid y_{1:k-1})$ is given by (8.4) and on substituting (8.4) in (8.3), we obtain (8.5).

$$p(x_k \mid y_{1:k-1}) = \int p(x_k \mid x_{k-1})p(x_{k-1} \mid y_{1:k-1})dx_{k-1} \tag{8.4}$$

$$p(x_k \mid y_{1:k}) = \frac{p(y_k \mid x_k)p(x_k \mid y_{1:k-1})}{p(y_k \mid y_{1:k-1})} \tag{8.5}$$

where $p(y_k \mid y_{1:k-1}) = \int p(y_k \mid x_k)p(x_k \mid y_{1:k-1})dx_k$ is the normalizing constant.

The variable $\hat{x}_k$ represents the mean of the desired posterior density function and is given by (8.6).

$$\hat{x}_k = \int x_k p(x_k \mid y_{1:k})dx_k \tag{8.6}$$

However, all of the above equations are conceptual problems and cannot be evaluated analytically as they involve calculations on infinite integrals [1, 3–5]. For a general distribution, the integrals cannot exhibit a closed-form solution and approximations are required, except when models are linear and have Gaussian characteristics. In EKF, all the involved distributions are Gaussian and are therefore completely represented by their mean and covariance. The mean and covariance of prior distribution $p(x_k \mid y_{1:k-1})$ are $\hat{x}_k^-$ and P_k^- (refer to Section 6.2.2), while those of previous posterior distribution $p(x_{k-1} \mid y_{1:k-1})$ are $\hat{x}_{k-1}^+$ and P_{k-1}^+; the posterior distribution $p(x_k \mid y_{1:k})$ is represented by $\hat{x}_k^+$ and P_k^+. However, if nonlinearities are severe, the EKF might diverge. To overcome the limitations of an EKF, other nonlinear filters are developed [5, 6].

Here suboptimal nonlinear filters are grouped into five main types [5, 6]:

- Analytical approximations;
- Numerical approximations;
- Gaussian sum or multiple model filters;
- Deterministic sampling filters; and
- Monte Carlo sampling filters.

Both the extended Kalman filter (EKF) along with its variations and iterated Kalman filters fall under the *analytical approximation* type of nonlinear filters. In this approach, the system and measurement models are linearized around a single best chosen point. But these filters suffer from divergence problems where the actual errors become inconsistent with the error covariance matrix (P_{xx}) approximated by the filter. This problem mainly arises when considerable levels of nonlinearities exist on the system, thus large linearization errors appear. Furthermore, the calculation of the Jacobians of the system and measurement models is a time-consuming process.

Grid-based methods [5] fall under the numerical approximation approach where solutions are evaluated at carefully selected grid points. The posterior density is approximated by a set of point values on a specified grid point. However, the grid needs to be dense to get a good approximation of the continuous state space. Therefore, as the dimensions of the state vector increase with time, there is an exponential increase in computational complexity. Moreover, state space must be predefined and therefore cannot be partitioned unevenly to give a higher resolution in higher-probability regions unless prior knowledge is used.

Gaussian filters are the filters that approximate the predictive conditional and posterior distributions as Gaussian densities. In these filters only the mean and the covariances of the distribution are propagated. Gaussian sum filters approximates the required posterior density function by a carefully weighted sum of Gaussian density functions. These filters are more accurate, especially for multimodal systems, but computationally expensive. Moreover, there is no automatic procedure to calculate the weights, mean, and covariances of the different Gaussian distributions.

Unscented Kalman filters (UKF) cover the *deterministic sampling filters* approach. A UKF deterministically samples a fixed number of minimal points (sigma points), which estimate the true mean and covariance of the Gaussian distribution according to the unscented transformation principle. These sigma points are then individually propagated through the true nonlinear system to accurately capture posterior mean and covariance. Using the unscented transformation principle, sigma points along with their respective weights are obtained, which satisfy the following conditions.

$$\sum_{i=0}^{p-1} w_i = 1 \tag{8.7}$$

$$\sum_{i=0}^{p-1} w_i X_i = \bar{x} \tag{8.8}$$

$$\sum_{i=0}^{p-1} w_i \left(X_i - \bar{x} \right)\left(X_i - \bar{x} \right)^T = P \tag{8.9}$$

where w_i are the weights of sigma points (x_i) and p is the number of sigma points. These sigma points have mean $\bar{x}$ with covariance P. However, when the nonlinearity is highly pronounced, even the best-fitting Gaussian distribution will be a poor approximation to the posterior distribution.

Particle filters (PFs) use the *Monte Carlo sampling* approach to obtain samples from the distributions. Particle filters are based on the principle of importance sampling and Monte Carlo. This algorithm was first used by Gordon, Salmond, and Smith in 1993 [11] (to the best of the author's knowledge) for non-Gaussian state models, and was called a Bayesian bootstrap filter. The same algorithm was proposed as a Monte Carlo filter along with a smoother by Kitagawa in 1996 [27], and as a condensation algorithm by Isard and Blake in 1996 [28]. This algorithm was first called particle filter (PF) by Carpenter, Clifford, and Fearhead in 1998 [29]. These PFs give an approximate solution to an exact model, rather than the optimal solution to an approximate model, which is the basis for the Kalman filter [7]. These suboptimal Bayesian filters approximate the posterior density function by random samples points called particles. Each of these particles has an assigned weight, and the required posterior density function is obtained as the summation of the product of these individual particles along with their respective weights [8]. The particle filter is based on the principle of importance sampling and Monte Carlo. Section 8.5

introduces the basic PF, which is a suboptimal filter used for cases where the dynamic models are nonlinear and the noises are non-Gaussian in nature [7, 8]. Before elaborating on PF principle, we will first explain the Monte Carlo principle, and the Importance Sampling method and Resampling methods.

8.2 The Monte Carlo Principle

The Monte Carlo (MC) principle is a stochastic sampling approach that analytically approximates intractable numerical integration problems [9, 10]. This principle uses a set of independent and identically distributed random samples to approximate the true integral representing the required PDF.

The MC principle states that if N independent, random, and equally weighted samples are drawn out from the required posterior distribution, then the required posterior PDF $p(x_{0:k} \mid y_{1:k})$ can be approximated by the average or a collection of these samples as given in (8.10):

$$p(x_{0:k} \mid y_{1:k}) = \frac{1}{N} \sum_{i=1}^{N} \delta(x_{0:k} - x_{0:k}^{(i)}) \tag{8.10}$$

where δ represents the Dirac delta function, $\{x_{0:k}^{(i)}, i = 1, 2 \ldots N\}$ is a drawn set of large N samples or particles with equal weights $\{w^{(i)} = 1/N, i = 1,2 \ldots N\}$. Now the expectations $I(f)$ of any function f can be approximated in summation form as given in (8.11) and (8.12):

$$I(f) = E[f(x_{0:k})] = \int f(x_{0:k}) p(x_{0:k} \mid y_{1:k}) dx_{0:k} \tag{8.11}$$

$$I(f) \approx \frac{1}{N} \sum_{i=1}^{N} f(x_{0:k}^{(i)}) \tag{8.12}$$

8.3 Importance Sampling Method

For the MC method, independent particles are required from the posterior density function. However, it is not always possible to draw out independent and uniformly distributed samples from the required density function $p(x_{0:k} \mid y_{1:k})$ as it may be multivariate or nonstandard. In these cases, the importance sampling method [11] is used. The importance sampling method states that if it is difficult to obtain samples directly from a distribution, then samples can be generated from the importance density function $q(x_{0:k} \mid y_{1:k})$, which is similar to the desired posterior distribution. But the weights of these generated samples need to be adjusted to represent the target function as closely as possible [1, 4, 12]. Starting from (8.10), and incorporating the importance density function into it, we obtain (8.13)–(8.15).

$$I(f) = \int f(x_{0:k}) \frac{p(x_{0:k} \mid y_{1:k})}{q(x_{0:k} \mid y_{1:k})} . q(x_{0:k} \mid y_{1:k}) dx_{0:k} \qquad (8.13)$$

$$I(f) = \int f(x_{0:k}) \frac{p(y_{1:k} \mid x_{0:k}) p(x_{0:k})}{p(y_{1:k}) q(x_{0:k} \mid y_{1:k})} . q(x_{0:k} \mid y_{1:k}) dx_{0:k} \qquad (8.14)$$

$$I(f) = \int f(x_{0:k}) \frac{w_k}{p(y_{1:k})} . q(x_{0:k} \mid y_{1:k}) . dx_{0:k} \qquad (8.15)$$

where unnormalized importance weights are given by $w_k = \dfrac{p(y_{1:k} \mid x_{0:k}) . p(x_{0:k})}{q(x_{0:k} \mid y_{1:k})}$.

Further, the unknown normalizing constant can be removed by performing the following steps.

$$I(f) = \frac{\int f(x_{0:k}) . q(x_{0:k} \mid y_{1:k}) . w_k . dx_{0:k}}{\int p(y_{1:k} \mid x_{0:k}) . p(x_{0:k}) . \dfrac{q(x_{0:k} \mid y_{1:k})}{q(x_{0:k} \mid y_{1:k})} dx_{0:k}} \qquad (8.16)$$

$$I(f) = \frac{\int f(x_{0:k}) . q(x_{0:k} \mid y_{1:k}) . w_k . dx_{0:k}}{\int w_k . q(x_{0:k} \mid y_{1:k}) . dx_{0:k}} \qquad (8.17)$$

$$I(f) = \frac{E[f(x_{0:k}) . w_k]}{E[w_k]} \qquad (8.18)$$

By drawing particles from this importance density function, we can approximate the expectation $I(f)$ with the Monte Carlo method [i.e.,(8.17) is approximated by (8.20)].

$$I(f) = E[f(x_{0:k})] \approx \frac{\dfrac{1}{N} \sum_{i=1}^{N} f(x_{0:k}^{(i)}) . w_k^{(i)}}{\dfrac{1}{N} \sum_{i=1}^{N} w_k^{(i)}} \qquad (8.19)$$

$$I = \sum_{i=1}^{N} \tilde{w}_k^{(i)} \cdot f(x_{0:k}^{(i)}) \qquad (8.20)$$

where $\tilde{w}_k^{(i)}$ are the normalized importance weights.

The exact form of the importance sampling distribution is a vital design issue since particles generated through this distribution will represent the required posterior distribution [14]. Furthermore, the weights of these generated particles are assigned according to the importance sampling distribution. The most popular and easiest option of importance density is the transition prior distribution as demonstrated by (8.21) [11].

$$q(x_k \mid x_{k-1}, y_{1:k}) = p(x_k \mid x_{k-1}) \qquad (8.21)$$

$$w_k \; \alpha \; w_{k-1} p(y_k \mid x_k) \qquad (8.22)$$

The use of the transition prior distribution as the importance sampling distribution greatly simplifies the particle weight calculations. The feasibility of readily obtaining particles from the transition prior density makes it an attractive choice. In this case, the weights depend only on the likelihood function. Consequently, after a few iterations, most of the samples have negligible weights as information coming from the latest measurement is completely ignored while drawing out particles. The most common ways to avoid this problem are to use a large number of particles or to implement a better proposal density function that is as close to the target distribution as possible or to use a resampling method [13].

8.4 Resampling Methods

Resampling eliminates particles with lower weights and multiplies particles with higher weights as shown in Figure 8.2. Resampling draws particles from the current distribution using the normalized weights as probabilities of selection. Furthermore, in resampling the weighted approximate density $\sum_{i=1}^{N} w_k^{(i)} \delta(x_k - x_k^{(i)})$ is converted to the unweighted density function $\sum_{i=1}^{N} (n_k^{(i)} / N) \delta(x_k - x_k^{(i)})$, as given in (8.24), so that most of the particles will eventually lie in higher posterior probability regions [30–33].

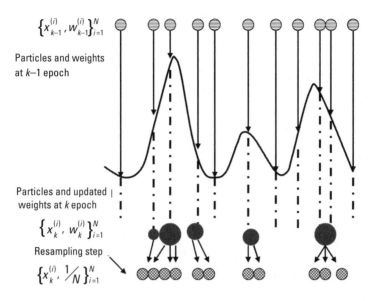

$\{x_{k-1}^{(i)}, w_{k-1}^{(i)}\}_{i=1}^{N}$

Particles and weights
at $k-1$ epoch

Particles and updated
weights at k epoch

$\{x_k^{(i)}, w_k^{(i)}\}_{i=1}^{N}$

Resampling step

$\{x_k^{(i)}, 1/N\}_{i=1}^{N}$

Figure 8.2   Resampling: removing low weight particles.

$$\sum_{i=1}^{N} w_k^{(i)} \delta(x_k - x_k^{(i)}) \Rightarrow \sum_{i=1}^{N} \frac{n_k^{(i)}}{N} \delta(x_k - x_k^{(i)}) \qquad (8.23)$$

$$\sum_{i=1}^{N} w_k^{(i)} \delta(x_k - x_k^{(i)}) = \frac{1}{N} \sum_{j=1}^{N} \delta(x_k - x_k^{*(j)}) \qquad (8.24)$$

Here $n_k^{(i)}$ is the number of replicates of particles $x_k^{(i)}$, which had a higher weight before the resampling step, and $\{x_k^{*(j)}\}_{j=1}^{N}$ is the new set of equally weighted particles. The resampling technique requires that the resampled density function should be close to the original distribution. Hence for any function $f(x)$, (8.25) should hold.

$$E\left[\left(\int f(x)p(x)dx - \int f(x)p(x)dx\right)^2\right] \xrightarrow{N \to \infty} (\qquad (8.25)$$

There are many different types of resampling methods [30], all of which are unbiased but differ in the variance aspect of the new unweighted density function being generated. The four main types of resampling techniques are discussed in Sections 8.4.1–8.4.4. For additional modified resampling methods on the combination of these fundamental four types please refer to [13].

8.4.1 Simple Random Resampling

The simple random resampling approach is the simplest approach, where a new particle set is generated by drawing N independent particles from the weighted old particles. The detailed steps of this method are as follows:

1. Generate a uniformly distributed random numbers such that $u_i \sim U(0,1)$.

2. Construct a cumulative density function of the particle weights and calculate $c_k = \sum_{j=1}^{k} w^{(j)}$.

3. In case u_i lies between c_{k-1} and c_k, only the particle with index i is chosen.

$$c_{k-1} < u_i < c_k \qquad (8.26)$$

4. Repeat this process until the total number of obtained particles is N.

Hence we observe that the number of offspring of the particle $x_k^{(i)}$ depends on how many times it is selected. Now, the collection of these particles with equal weights approximates the posterior density function. The variance of the simple random sampling method is equal to the variance of $g(x)$ divided by total number of particles [29, 30].

8.4.2 Systematic Resampling (SR)

The SR algorithm [29] is very similar to the simple random resampling method explained in Section 8.4.1. However unlike simple random resampling, where each random number u_i is independent from the uniform distribution for $i = 1, \ldots, N$, in this case N ordered random numbers are drawn by using (8.27).

$$u_i = \frac{(i-1) + u_1}{N} \qquad (8.27)$$

where $u_i \sim U[0,1]$ and N is the total number of resampled particles. In systematic resampling, the generated particles are dependent on each other and therefore correct variance analysis cannot be conducted. However, variance of this method [32] is assumed to be less than stratified resampling, as particles generated by this method have the least discrepancy since it ensures that each sample is selected only once, and hence no two samples are the same in this method.

8.4.3 Stratified Resampling

In stratified resampling [13], the interval $(0, 1)$ is divided into N strata of equal size $\left[\dfrac{i-1}{N}, \dfrac{i}{N}\right)$ for $i = 1, \ldots, N$ and then a single random sample is drawn from each of the intervals according to (8.28).

$$u_i = \frac{(i-1) + u_i}{N} \tag{8.28}$$

where $u_i \sim U[0,1]$.

Here particles are more uniformly distributed than in the simple random resampling method. Therefore stratified variance is less than that of the simple random resampling method in Section 8.4.1.

8.4.4 Residual Resampling

Residual resampling [30, 33] is a different approach than the rest of the above resampling methods as it sets restrictions instead of sampling methods. This method is composed of two steps. In the first step, the number of replications of the particles (before the resampling step) is calculated. This can be achieved by taking the integer part of Nw_i for each particle. However, since the first step does not guarantee that the number of resampled particles remains N, the residual N_r is computed in the second step. These residual particles are randomly selected with replacement from the original particles using any one of the previous resampling methods. The residual resampling method's variance is also less than that of the simple random resampling method. At this point, we have sufficient background to start our discussion on basic PF.

8.5 Basic Particle Filters

PFs construct a point mass representation of a state vector by a large number of random particles that explore the state space [5, 6]. PFs are based on the Sequential Monte Carlo estimation (SMC) method, which is the integration of the Monte Carlo and importance sampling methods [7, 8, 14]. In PF, the posterior density function is represented by a collection of weighted particles that are generated by the Monte Carlo principle. Unlike an EKF, these particles are then individually propagated through the true nonlinear system. Particle filters allow for a complete representation of the posterior distribution of the states, so that any statistical estimates, such as mean and covariance, can be easily computed. These filters are capable of dealing with any nonlinearity in the models. Some examples of PF implementation for land vehicles are provided in [15–17]. Using this large number of particles significantly increases the computational load of the filter, and, hence, modified versions of PFs have been developed and are covered in Sections 8.6 and 8.7.

8.6 Types of Particle Filters

One prominent type of particle filter is the extended or unscented particle filter.

8.6.1 Extended Particle Filter (EPF) and Unscented Particle Filter (UPF)

This section presents the methodology and implementation of EPFs to eliminate the shortcomings of the basic PFs. In addition, to validate the developed EPF algorithm, its performance is compared to that of the EKF for real field test datasets in a variety of scenarios. In basic PFs, the additional knowledge coming from the latest measurement is not considered in the importance density function. This choice of importance distribution simplifies the calculation of importance weights. However, problems occur in cases where the likelihood is too narrow in comparison to the transition prior density function, and, therefore, the particles should be moved to high likelihood regions [12].

8.6.1.1 Methodology of EPF/UPF

One method of moving particles to a higher likelihood region is to use a better importance distribution. This can be accomplished by choosing an importance density that is conditioned on the latest measurement. Kalman-based filters, like an EKF or UKF, incorporate the latest measurement into the updated posterior state [12]. Therefore, if an EKF or UKF is used to generate the importance distribution, the latest measurement can be incorporated into the distribution. In an EPF or UPF, incorporation of the most current observation into the state is realized through the local linearization of the state vector by an EKF [1, 4, 12, 21–23]. In the local linearization technique promoted by

Doucet [1], each particle is drawn from the local Gaussian approximation of the optimal importance distribution that is conditioned both on the current state and the latest measurement [i.e., (8.29)].

$$q(x_k \mid x_{k-1}, y_k)_{EKF/UKF} \approx q_N(x_k \mid y_{1:k}) \tag{8.29}$$

where q_N denotes the Gaussian approximation of the optimal importance density function. This choice of the importance density function can be expanded as shown in (8.30), which modifies the importance weights to be represented by (8.33).

$$q(x_k \mid x_{k-1}, y_k)_{EKF/UKF} = \frac{p(y_k \mid x_k) p(x_k \mid x_{k-1})}{p(y_k \mid x_{k-1})} \tag{8.30}$$

$$w_k \; \alpha \; \frac{p(y_k \mid x_k) p(x_k \mid x_{k-1})}{q(x_k \mid x_{k-1}, y_k)_{EKF/UKF}} \tag{8.31}$$

$$w_k \; \alpha \; \frac{p(y_k \mid x_k) p(x_k \mid x_{k-1})}{p(y_k \mid x_k) p(x_k \mid x_{k-1}) \Big/ p(y_k \mid x_{k-1})} \tag{8.32}$$

$$w_k \; \alpha \; p(y_k \mid x_{k-1}) \tag{8.33}$$

A tractable way of generating the Gaussian approximated importance density function in the PF framework is to use a separate EKF/UKF to generate and propagate a Gaussian distribution for each particle [12]; that is,

for $i = 1, \ldots, N$

$$q(x_k^{(i)} \mid x_{k-1}^{(i)}, y_k)_{EKF/UKF} = N(x_k^{(i)}, P_k^{(i)}) \tag{8.34}$$

where $x_k^{(i)}$ and $P_k^{(i)}$ are the mean and variance of the ith particle generated by the EKF/UKF algorithm. Once all the particles have been updated by the EKF/UKF, particles are redrawn from this generated importance distribution. These particles are now conditioned on all the states and measurements available until the current time epoch. In our particular application of a land vehicle navigation system, the UKF and EKF gave a similar performance [25]. It was further observed that the processing time involved in the UKF was much larger than that of the EKF, and hence we only show the implementation of the EPF and not that of the UPF.

The flowchart for the EPF algorithm is shown in Figure 8.3, which shows three main steps (i.e., importance sampling, updating weights of the particles, and the resampling step). The modified state vector, which is very similar to the EKF state vector, is given in Figure 8.4 for both loosely and tightly coupled approaches. A loosely coupled state vector consists of position errors, velocity errors, attitude errors, and bias and scale factor errors for both accelerometers and gysoscopes. For a tightly coupled state vector, all these error states along with the clock bias and clock drift errors are included.

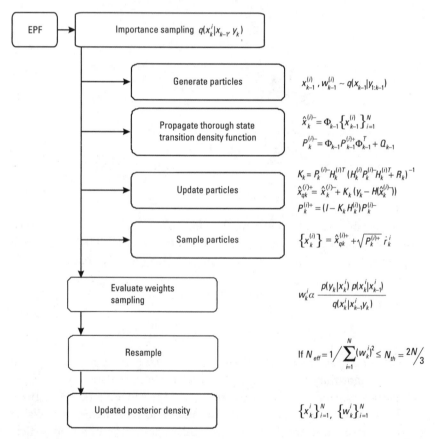

Figure 8.3 Flowchart of an extended particle filter.

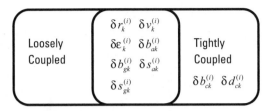

Figure 8.4   Loosely coupled versus tightly coupled.

The basic difference in the EPF state vector as compared to the EKF state vector is that instead of a single value for each state variable, a cluster of particles is used. Like the EKF, the EPF state vector models sensor noises along with sensor errors to compensate for the effect of noises, biases, and scale factor errors on the INS measurements.

Any other land vehicle constraints can be implemented in the EPF similar to the EKF except for the fact that a cluster of states will be used instead of a single state. For example, an NHC can be incorporated as follows for N particles.

$$v_y^b = \left\{ v_y^{b(i)} \right\} \approx 0 \tag{8.35}$$

$$v_z^b = \left\{ v_z^{b(i)} \right\} \approx 0 \tag{8.36}$$

where $i = 1, \ldots, N$ are the individual particles of each state variable in EPF. A comparison of EPF results with EKF is provided in subsection 8.6.1.2.

8.6.1.2   An Example of EPF Results for a Land Vehicle

The EPF performance is shown in this example for a field dataset. The dataset was collected in August 2004 using a CIU. In addition to the CIU, the test vehicle also carried NovAtel OEM4 GPS receivers and a tactical grade inertial unit for the reference trajectory.

Trajectory 1 is provided in Figure 8.5 and its velocity profile is shown in Figure 8.6. Figure 8.6 also shows the periods when the vehicle stopped and the location of simulated GPS signal outages. The performance of the EPF will be evaluated during the signal outages when the EPF is running only in the prediction mode and its output will be compared to the most commonly used EKF for the same conditions. For trajectory 1, six simulated GPS outages of 60 seconds exist both at the high and no dynamic portions of the trajectory (see Figure 8.6).

Figure 8.5 Field test trajectory 1 of CIU.

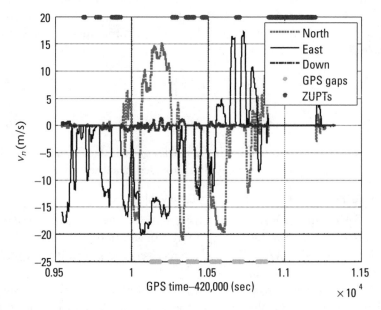

Figure 8.6 CIU trajectory 1 velocity profile.

A high dynamic region occurs at either turns or where vehicle velocities are substantial. No dynamic region refers to the periods of trajectory when the vehicle is stationary or when ZUPTs are encountered. GPS signal outages 1, 4, and 6 lie at high dynamic regions, while the rest of the outages lay in partial high and partial no dynamic areas. After the explanation of the trajectory, let us move ahead and discuss the format for the analysis. The comparison is

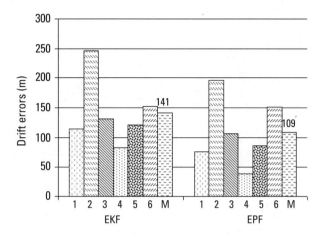

Figure 8.7 CIU field tests results for trajectory 1.

made when the navigation is purely MEMS inertial sensors based. For that, no aiding was provided by simulating the GPS signal outages as mentioned earlier. The maximum 3D position errors were computed based on the reference trajectory values during those periods. The 3D errors are referred to as square root of the sum of squares of x, y, and z position errors, and provided for both the EKF and EPF in Figure 8.7.

In this Figure 8.7, the x-axis shows the six outages along with the mean drift error. From this figure, we can clearly observe that there is a 28% improvement in the EPF performance as compared to the EKF. The main reason for this improvement is better performance during higher dynamic regions, such as sharp 90° turns and the freedom to use any suitable distribution instead of Gaussian distribution. In all of these results, gamma distribution is being used for the prior density function to give more weights to the current likelihood function. Figure 8.8 illustrates the results obtained when posterior distribution is Gaussian and non-Gaussian for the EPF case. This figure also shows that a similar result as that of the EKF is obtained if the posterior distribution (in case of an EPF) is Gaussian.

In this particular example, there are five 60-sec GPS outages and the performance of each filter is evaluated during this time. Hence the main advantage of an EPF (in this implementation) lies in the freedom of the particle filter to use any kind of distribution; the distribution choice is not limited to a Gaussian distribution.

Other types of particle filters are briefly given in Sections 8.6.2–8.6.5.

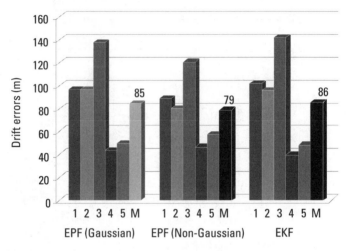

Figure 8.8   EPF performance with Gaussian and non-Gaussian distributions.

8.6.2   Rao-Blackwellized Particle Filter (RBPF)

Basic particle filters require a large number of particles to accurately approximate the posterior density function. One way of reducing this computational cost is to divide the state vector into two components, such that only a small part of the state vector is sampled by the PF. In a Rao-Blackwellized particle filter, the state vector x_k is divided into two components, namely the linear part x_k^l and nonlinear part x_k^{nl} [26].

$$p(x_{0:k}, y_{1:k}) = p(x_{0:k}^{nl}, x_{0:k}^l \mid y_{1:k}) = p(x_{0:k}^l \mid y_{1:k}, x_{0:k}^{nl}) p(x_{0:k}^{nl} \mid y_{1:k}) \quad (8.37)$$

$$p(x_{0:k}, y_{1:k}) = p(x_{0:k}^l \mid y_{1:k}, x_{0:k}^{nl}) p(x_{0:k}^{nl} \mid y_{1:k}) \quad (8.38)$$

On applying the Bayes theorem, the linear part is marginalized out such that the linear part $p(x_{0:k}^l \mid y_{1:k}, x_{0:k}^{nl})$ is exactly evaluated by the KF while for the nonlinear part $p(x_{0:k}^{nl} \mid y_{1:k})$, the PF is implemented. Hence this marginalized posterior distribution will require a smaller number of particles and will have a smaller variance [7, 8, 18, 20]. However the marginalized system will no longer be Markovian.

8.6.3   Likelihood Particle Filter (LPF)

In an SIS/R PF, the transition density is taken as the proposal density function. In cases where the likelihood is too narrow or peaked, very few particles will lie in a high-likelihood region. Therefore most of the particles will have a negligible weight, which gives an inaccurate posterior density estimate. A possible way to reduce this effect is to design a better proposal density function. One of the ways is to draw out particles from the likelihood function and then weigh the particles according to the transition density function [6]. The particle filter where the likelihood function is used as the importance density function is called the likelihood particle filter (LPF).

8.6.4 Regularized Particle Filter (RPF)

The regularized particle filter was developed to remove the additional problem caused by resampling in the SIR filter (i.e., particle impoverishment or reduction in particle diversity). A regularized particle filter is identical to the SIR filter except in the resampling stage. An RPF resamples from a continuous approximation of the posterior density instead of a discrete approximation, so that the particles do not collapse to a single sample in cases where model noises are much less [8, 18–20]. In an RPF, additional noise terms are added in the form of the optimized kernels applied on each particle. RPF samples are drawn from the approximation:

$$p(x_k \mid y_{1:k}) \approx \sum_{i=1}^{N} w_k^i K_h(x_k - x_k^i) \tag{8.39}$$

where $K_h(x) = \frac{1}{h^{n_x}} K(\frac{x}{h})$ is the rescaled kernel density $K(.)$, h is the kernel bandwidth, and n_x is the dimension of the state vector x with normalized weights w_k^i. Kernel density is a symmetric probability density function with properties $\int_{-\infty}^{\infty} K(x)dx = 1$ and $K(-x) = K(x)$ for all x. The main advantage over an SIR filter is that RPF performance is better in cases where severe sample impoverishment occurs due to less process noise. However, these samples no longer asymptotically approximate samples of the posterior distribution.

8.6.5 Gaussian Particle Filter (GPF) and Gaussian Sum Particle Filter (GSPF)

Gaussian filters (GFs) approximate the predictive conditional distribution and posterior distribution as Gaussian densities [33], while Gaussian sum filters (GSFs) approximate the posterior distribution as a Gaussian mixture. However, their performance degrades when significant nonlinearities exist in the system. To overcome these problems, new filters called Gaussian particle filters [(GPF)—based on the GFs principle] and Gaussian sum particle filters [(GSPF)—based on the GSFs principle] were suggested [33, 34].

In a GPF, the posterior mean and covariance of the state vector is approximated using the importance sampling method. Assume the previous posterior density function $p(x_{k-1} \mid y_{1:k-1})$ is approximated by a Gaussian distribution $N(x_{k-1}; \bar{x}_{k-1}; P_{k-1})$, where $\bar{x}_{k-1}$ is the mean and P_{k-1} is the covariance of the Gaussian distribution. Samples $x_{k-1}^{(i)}$ are drawn from this distribution and are passed through the system model to obtain the predicted samples $x_k^{(i)}$. Hence the prior density is also a Gaussian distribution with mean $\bar{x}_k = \sum_{i=1}^{N} w_{k-1}^{(i)} x_k^{(i)}$ and covariance $P_k = \sum_{i=1}^{N} w_{k-1}^{(i)} (x_k^{(i)} - \bar{x}_k)(x_k^{(i)} - \bar{x}_k)^T$. On receiving a new y_k measurement, the weights of these sampled particles are updated. If the selected importance density function is equal to the prior density, weights are proportional to the likelihood function only. Now the collection of particles along with their weights approximates the posterior density function. No resampling step is implemented in this type of filter as the posterior distribution is approximated by the mean and covariance like in an EKF. The removal of the resampling step reduces the computational cost and allows the parallel processing of the algorithm. The Gaussian sum particle filter (GSPF) is very similar to the GPF [33, 34]. The only difference is that instead of approximating the predictive and filtering distribution (posterior distribution) by a Gaussian distribution, it is approximated by a mixture of Gaussian distributions. This technique is very useful in cases where the posterior distribution is multimodal. However, it is not always easy to accurately obtain a mixture of Gaussian distributions with reasonable weights.

A hybrid extended particle filter is a novel filter proposed by us. This filter combines the advantage of a Kalman filter for the linear portion of the trajectory with that of an EPF for the nonlinear portion of the trajectory. This filter provides an accurate navigation solution along with a higher processing speed than any other form of particle filter. Unlike an RBPF, it doesn't divide the state vector into two portions but alternates between a KF and EPF depending on vehicle dynamics.

8.7 Hybrid Extended Particle Filter (HEPF)

An EPF gives a better performance in cases where GPS signal outages occur at high dynamic areas as exemplified in the previous section. During ZUPTs, the vehicle is in a stationary condition (i.e., there is no motion and therefore there is no nonlinearity in the position estimation). Equations (8.40)–(8.42) are the components of mechanization equations that are nonlinear during vehicle motion. However, when the vehicle is stationary all the velocity components should ideally be 0 and (8.40)–(8.42) reduce to (8.43)–(8.45), which are no longer nonlinear in nature.

$$\varphi(t_{k+1}) = \varphi(t_k) + \frac{1}{2}\frac{[V^n(t_{k+1}) + V^n(t_k)]}{R + h}\Delta t \tag{8.40}$$

$$\lambda(t_{k+1}) = \lambda(t_k) + \frac{1}{2}\frac{[V^e(t_{k+1}) + V^e(t_k)]}{(R + h)\cos\varphi}\Delta t \tag{8.41}$$

$$h(t_{k+1}) = h(t_k) + \frac{1}{2}[V^d(t_{k+1}) + V^d(t_k)]\Delta t \tag{8.42}$$

$$\varphi(t_{k+1}) = \varphi(t_k) \tag{8.43}$$

$$\lambda(t_{k+1}) = \lambda(t_k) \tag{8.44}$$

$$h(t_{k+1}) = h(t_k) \tag{8.45}$$

Under these constant conditions, an EKF is the preferred option in comparison to an EPF. Based on this principle, and to overcome the computational load limitation of an EPF, a hybrid extended particle filter is proposed [24].

An adaptive subroutine is developed based on the zero velocity periods to switch between an EPF and EKF depending on the vehicle dynamic. In cases where the dynamic is high, an EPF is implemented while in cases where it is less (i.e., a zero velocity condition exists), an EKF is incorporated. In this novel filter structure, an algorithm has been realized to detect ZUPTs conditions depending on the raw accelerometer signals. The first step for implementing an HEPF is to detect ZUPTs as discussed in Section 8.7.1.

8.7.1 Zero Velocity Condition Detection Algorithm

As mentioned earlier, ZUPTs occur when the land vehicle is stationary and represents no dynamic region in the vehicle trajectory. The velocity errors at this time are due to short-term random disturbances, which can be reduced by averaging the measurements. When the vehicle stops, corrections are made to all error states modeled in the filter that compensates or reduces the errors.

One method that can be used to detect the ZUPTs using raw inertial signals is explained in Section 8.7.1.1.

8.7.1.1 ZUPTs Detection Code

1. If all the velocities in the three directions are less than 0.5 m/s during the first 150 accelerometers data records, the standard deviation of the raw accelerometer signal values within these data records becomes the threshold. In case velocities are not less than 0.5 m/s, the designed algorithm looks for data where the condition is met. The value 0.5 m/s has been selected on examining various field datasets and is accurate enough to distinguish the static portion from the dynamic portion. Too large a value can misjudge the dynamic portion as a static portion, while too small a value can misinterpret static to be a dynamic region. Henceforth, this value is an essential tuning component for a reasonable ZUPTs detection technique.

2. Next, a sliding window of a range of 7–25 seconds is chosen, and the standard deviations of the raw accelerometer signals within this window are compared to the threshold value. This window size is another important tuning parameter. This algorithm has been able to accurately detect ZUPTs with a sliding window size as small as 7 sec. It is not always beneficial to frequently switch filters since few epochs are required by the newly generated particles to converge to the ideal solution. Generally, a window size of 10–15 seconds is considered optimal.

3. Once the threshold value is obtained, the standard deviation of the raw accelerometer signals within the selected window is compared with this threshold. If these accelerometer values are less than the threshold value, then ZUPTs has been detected.

$$std(acc) <= acc_{Thres}, \qquad \text{ZUPT detected} \qquad (8.46)$$

where acc_{Thres} is the threshold standard deviation obtained from step 1.

Let us now look at the complete algorithm for the implementation of a hybrid extended particle filter.

8.7.2 Algorithm of the Hybrid Extended Particle Filter

An HEPF combines the advantages of both an EKF and EPF by alternating between them based on the vehicle dynamic identified by ZUPTs to increase the processing speed and accuracy of the navigation solution.

8.7.2.1 Pseudocode for HEPF

1. Set initial epoch generate N samples $\{x_0^i\}_{i=1}^N$ from the prior distribution $p(x_0)$.

2. Calculate the threshold required to detect ZUPTs. Select a timing window and based on Section 8.7.2, identify ZUPTs.

3. If ZUPTs is detected, implement EKF; otherwise, continue implementing an EPF.

EPF Implementation

1. *Importance sampling step:* Sample particles from the previous importance density function [i.e., $\{x_{k-1}^{(i)}, w_{k-1}^{(i)}\}_{i=1}^N \rightarrow q(x_{k-1} \mid y_{1:k-1})$. This is executed by the following steps.

   For $i = 1,\ldots, N$

   - Calculate the Jacobians $\Phi_{k-1}^{(i)}$ and $H_k^{(i)}$ of the process and measurement models.

   - Propagate the particles through the state transition density function and estimate the covariance of the particles.

   - Update the particles along with their weights.

2. Sample N particles from the updated importance distribution.

3. For $i = 1,\ldots, N$, evaluate the importance weights of each particle.

4. For each particle $i = 1,\ldots, N$, normalize the weights.

5. Compute the effective weights and threshold.

6. If $N_{eff} > N_{th}$, particles do not change, or else higher weight particles are repeated (i.e., the resampling step is performed).

EKF Implementation

1. Calculate the mean vector of dimension 21×1 and its covariance matrix.

2. Calculate the Jacobians of the process and measurement models.

3. Propagate the mean value through the state transition density function and estimate its covariance.

4. Update the mean state vector along with its covariance.

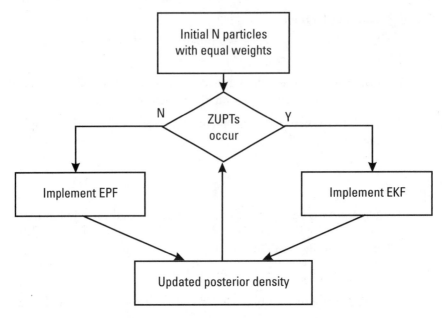

Figure 8.9   Algorithm of a hybrid extended particle filter.

5.   Check for ZUPTs; if detected repeat EKF implementation, otherwise, generate particles for the EPF and implement the EPF portion of the algorithm.

A simplified version of the HEPF algorithm is provided in Figure 8.9.

8.7.3 HEPF Results

An HEPF can be implemented using either loosely or tightly coupled integration strategies. Sections 8.7.3.1–8.7.3.2 are examples used to show the effectiveness of HEPF for both loosely coupled integration schemes.

8.7.3.1 An Example of HEPF with Loosely Coupled Integration

This example will provide the results of two different MEMS-grade IMUs, which are the CIU and the Motion Pak II units with gyro drifts of 1°/sec and 0.5°/sec, respectively. Both datasets were collected in Calgary, Canada, and included a GPS receiver and a tactical-grade inertial system onboard the test vehicle for reference in addition to the test MEMS IMUs. The velocity profile for CIU trajectory 1 is shown in Figure 8.6 while the velocity profile for Motion Pak II is illustrated in Figure 8.10. For all trajectories, short periods of GPS outages are carefully selected to cover different vehicle dynamics. Figure 8.10 illustrates that outage 1 lies in a partial no dynamic and partial high dynamic region while the rest of the outages are in the higher dynamic regions of the trajectory.

The loosely coupled integration results using an EKF and an HEPF are provided below in Figure 8.11 for CIU. The results are for GPS signal outage periods to estimate the usability of the two filters. However, both ZUPTs and nonholonomic constraints are applied as these two constraints are going to be part of most land vehicle trajectories.

We can clearly see an improvement of 34.3% in the HEPF results as compared to that of the EKF. The next example is provided to compare the advantages of an HEPF for shorter GPS signal outage periods. For this proof, we simulated 30-sec outages, ran the two filters, and compared the results in Figure 8.12.

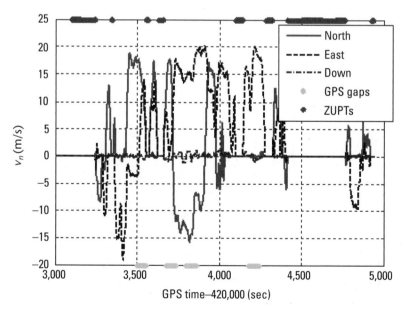

Figure 8.10 Motion Pak II trajectory 1 velocity profile.

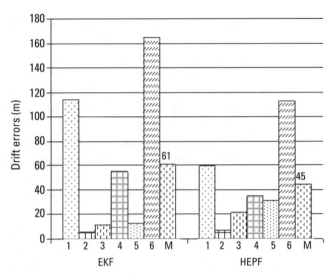

Figure 8.11   EKF and HEPF results for CIU 60-sec outages.

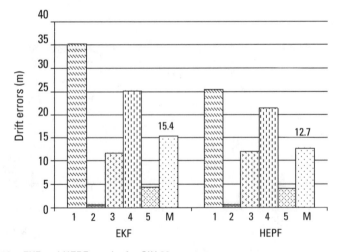

Figure 8.12   EKF and HEPF results for CIU 30-sec outages.

We can clearly see that the improvement is reduced to 23.3% from 34.3% when the duration of outages is reduced to 30 sec instead of 60 sec. Therefore, it can be deduced that the longer the outages are, the better the performance

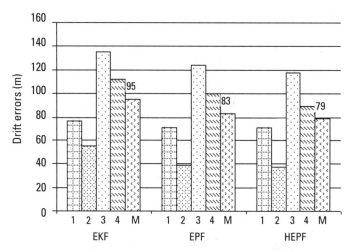

Figure 8.13   EKF, EPF, and HEPF results for Motion Pak II.

of an HEPF. The example provided below shows the comparison among different filters for MotionPak II, which is of significantly better quality than the CIU (Figure 8.13).

All of the above results are for a loosely coupled integration scheme that requires at least 4 GPS satellites to produce position and velocity information. However, in urban areas this is generally not feasible because of the lack of a direct line of sight to four or more satellites. For these cases, a tightly coupled integration scheme is very useful. In Section 8.7.3.2 we show the results obtained for a tightly coupled integration scheme.

8.7.3.2 An Example of HEPF with Tightly Coupled Integration

To compare the performance of different filters, 60-sec periods with less than 4 satellites were simulated. The IMU position errors during these periods are obtained by comparing the corresponding solution to the reference trajectory as described earlier. For the trajectory provided in Figure 8.14, four 60-sec GPS signal degradation periods were simulated. The first 2 periods mimic the situation when there is no direct line of sight to the GPS satellites (i.e., 0 satellites available). The third degradation period had signals from 3 satellites while the last degradation period shows the situation where the GPS receiver has a direct line of sight to only 1 satellite. Figure 8.15 provides the comparison for an EKF and HEPF when tightly coupled integration strategies were implemented.

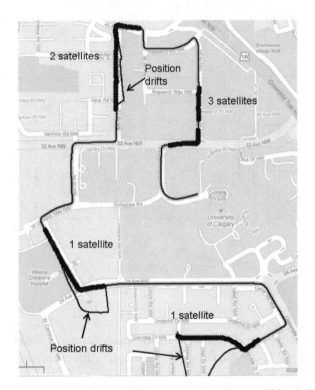

Figure 8.14 Tightly coupled integration CIU trajectory for different GPS satellite availability.

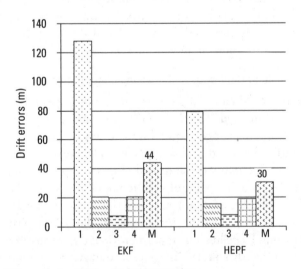

Figure 8.15 CIU results for tightly coupled integration.

From Figure 8.15 we can clearly see that an HEPF gives a 44.8% better performance than that of an EKF in the tightly coupled integration scheme. This is mainly due to degradation period 1, which lies in a high dynamic region. Moreover, we also conclude that the performance of CIU is much better when INS and GPS datasets are integrated by a tightly coupled scheme rather than a loosely coupled scheme.

Now let's investigate the performance of particle filters for partial sensor configurations as discussed in Chapter 4.

8.7.4 Partial Sensor Configuration

For a land vehicle, a reduced number of applicable sensors are rather than the necessary 3 gyroscopes and 3 accelerometers, as vehicle motion is mainly constrained to two dimensions.

8.7.4.1 An Example of Partial Sensor Configuration

The results for loosely coupled partial sensor configurations are divided into three main subcategories. The first portion presents the position drift errors for an EPF and an HEPF with a 1G3A configuration, while the second part shows the results with a 1G2A sensor configuration (Figure 8.16). The last section demonstrates the results for full IMU configuration. 1G3A results indicate a 12.85% improvement over an EPF.

The improvement was even better when a 1G2A sensor configuration was used. The HEPF performed 23% better than the EPF when the vertical

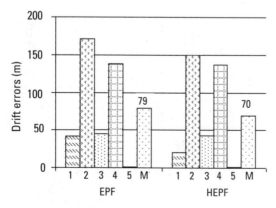

Figure 8.16 EPF and HEPF results for 1G3A configuration.

accelerometer was replaced by an ideal signal of -9.81 m/s^2 as illustrated in Figure 8.17.

The result of using full sensors is given in Figure 8.18, which shows an average accuracy improvement of 18.7% over the EPF. It is also important to note that results for full IMU are significantly better than any partial IMU.

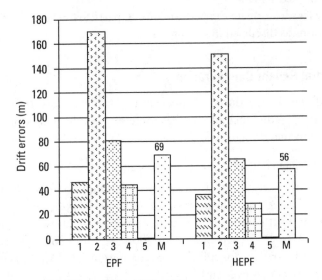

Figure 8.17   EPF and HEPF results for 1G2A configuration.

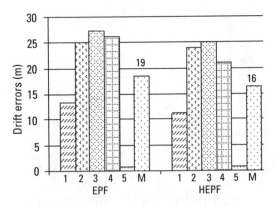

Figure 8.18   EPF and HEPF results for full IMU configuration.

In the end we establish that particle filters, namely, the EPF and HEPF filters, give a much better performance than an EKF, especially when the GPS outages are at high dynamic regions (i.e., sharp turns, and so forth). We have seen an improvement up to 44% for the HEPF in comparison to the EKF for a tightly coupled integration scheme. Furthermore, it is established that for a land vehicle, a partial sensor configuration of 1G2A is most economical to use and gives a reasonable navigation solution.

This completes our efforts for conveying sufficient knowledge to implement integrated navigation system based on low-cost MEMS sensors.

References

[1]   Doucet, A., "On Sequential Simulation-Based Methods for Bayesian Filtering," Tech. Rep. CUED/F-INFENG/TR.310, Department of Engineering, University of Cambridge, 1998.

[2]   Murphy, K., "Dynamic Bayesian Networks: Representation, Inference and Learning," Ph.D. thesis, U.C. Berkeley, 2002.

[3]   Punskaya, E., "Sequential Monte Carlo Methods for Digital Communication," Ph.D. thesis, University of Cambridge, 2003.

[4]   Doucet, A., Freitas, N., and Gordon, N., *Sequential Monte Carlo Methods in Practice*, New York: Springer Verlag, 2001.

[5]   Maksell, S., and Gordon, N., "A Tutorial on Particle Filters for On-Line Nonlinear/Non-Gaussian Bayesian Tracking," *IEE Target Tracking: Algorithms and Applications*, Vol. 2, 2001, pp. 1–15.

[6]   Arulampalan, S., et al., "A Tutorial of Particle Filters for Online Nonlinear/Non-Gaussian Bayesian Tracking," *IEEE Transactions on Signal Processing*, Vol. 50, No. 2, 2002, pp. 174–188.

[7]   Nordlund, P.J., "Sequential Monte Carlo Filters and Integrated Navigation," thesis no. 945, Linköping Univ., Linköping, Sweden, 2002.

[8]   Ristic, B., Arulampalan, S., and Gordon, N., *Beyond the Kalman Filter: Particle Filters for Tracking Applications*, Norwood, MA: Artech House, 2004.

[9]   Handschin, J.E., "Monte Carlo Techniques for Prediction and Filtering of Non-Linear Stochastic Processes," *Automatica*, Vol. 6, 1970, pp. 555–563.

[10]   Zaritskii, V.S., Svetnik, V.B., and Shimelevich, L.I., "Monte Carlo Technique in Problems of Optimal Data Processing," *Autom. Remote Control*, Vol. 12, 1975, pp. 95–103.

[11]   Gordon, N., Salmond, D.J., and Smith, A.F.M., "Novel Approach to Nonlinear/Non-Gaussian Bayesian State Estimation," *Proc. Radar and Signal Processing*, Vol. 140, 1993, pp. 107–113.

[12] Doucet, A., Godsill, S.J., and Andrieu, C., "On Sequential Simulation-Based Methods for Bayesian Filtering," *Statistics and Computing*, Vol. 10, No. 3, 2000, pp. 197–208.

[13] Hol, J.D., "Resampling in Particle Filters," Ph.D. thesis, Linköping University, Linkoping: Sweden, 2004.

[14] Alspach, D.L., and Sorenson, H.W., "Nonlinear Bayesian Estimation Using Gaussian Sum Approximations," *IEEE Transactions on Automatic Control*, Vol. AC-17, 1972, pp. 439–448.

[15] Giremus, A., Tourneret, J.Y., and Djuric, P.M., "An Improved Regularised Particle Filter for GPS/INS Integration," *The Sixth IEEE International Workshop on Signal Processing Advances in Wireless Communications (SPAWC)*, New York, 2005, pp. 1013–1017.

[16] Yi, Y., and Grejner-Brzezinska, D.A., "Tightly-Coupled GPS/INS Integration Using Unscented Kalman Filter and Particle Filter," *Proceedings of the Institute of Navigation— 19th International Technical Meeting of the Satellite Division (ION GNSS 2006)*, Vol. 4, Fort Worth, TX, 2006, pp. 2182–2191.

[17] Georgy, J., Iqbal, U., and Bayoumi, M., "Reduced Inertial Sensor System (RISS)/GPS Integration Using Particle Filtering for Land Vehicles," *Proceedings of the 21st International Technical Meeting of the Satellite Division of the Institute of Navigation (ION GNSS 2008)*, Savannah, GA, 2008.

[18] Chen, Z., "Bayesian Filtering: From Kalman Filters to Particle Filters and Beyond," technical report, Adaptive Systems Lab., McMaster University, Hamilton, ON, Canada, 2003.

[19] Kong, A., Liu, J.S., and Wong, W.H., "Sequential Imputations and Bayesian Missing Data Problems," *J. Amer. Statist. Assoc.*, Vol. 89, 1994, pp. 278–288.

[20] Smith, A., et al., *Sequential Monte Carlo Methods in Practice, First Edition*, New York: Springer Verlag, 2001.

[21] Aggarwal, P., Gu, D., and El-Sheimy, N., "Extended Particle Filter (EPF) for Land Vehicle Navigation Applications," *International Global Navigation Satellite Systems (IGNSS 2007)*, Sydney, Australia, 2007.

[22] Aggarwal, P., and El-Sheimy, N., "Extended Particle Filter (EPF) for Land and Airborne Applications," submitted to *Measurement Science and Technology*, 2008.

[23] Merwe, R., et al., "The Unscented Particle Filter," *Neural Information Processing Systems*, Vol. 13, 2000.

[24] Aggarwal, P., Syed, Z.F., and El-Sheimy, N., "Hybrid Extended Particle Filter (HEPF) for INS/GPS Integrated System," accepted for publication in *Measurement Science and Technology*, 2008.

[25] Shin, E.-H., "Estimation Techniques for Low-Cost Inertial Navigation," Ph.D. thesis, Department of Geomatics Engineering, University of Calgary, UCGE Report No. 20219, 2005.

[26]   Doucet, A., et al., "Rao-Blackwellised Particle Filtering for Dynamic Bayesian Networks," *Proc. UAI2000*, 2000, pp. 176–183.

[27]   Kitagawa, G., "Monte Carlo Filter and Smoother for Non-Gaussian Nonlinear State Space Models," *Journal of Computational and Graphical Statistics*, Vol. 5, 1996, pp. 1–25.

[28]   Isard, M., and Blake, A., "CONDENSATION Conditional Density Propagation for Visual Tracking," *International Journal of Computer Vision*, Vol. 29, 1998, pp. 5–28.

[29]   Carpenter, J., Clifford, P., and Fearnhead, P., "Improved Particle Filter for Nonlinear Problems," *IEE Proceedings on Radar and Sonar Navigation*, Vol. 146, 1998, pp. 2–7.

[30]   Liu, J.S., and Chen, R., "Blind Deconvolution via Sequential Imputations," *J. Amer. Statist. Assoc.*, Vol. 90, 1995, pp. 567–576.

[31]   Liu, J.S., Chen, R., and Wong, W.H., "Rejection Control and Sequential Importance Sampling," *J. Amer. Statist. Assoc.*, Vol. 93, 1998, pp. 1022–1031.

[32]   Bolic, M., Djuric, P.M., and Hong, S., "New Resampling Algorithms for Particle Filters," *Proc. ICASSP*, Vol. 2, 2003, pp. 589–592.

[33]   Kotecha, J.H., and Djuric, P.M., "Gaussian Sum Particle Filtering," *IEEE Transactions on Signal Processing*, Vol. 51, 2003, pp. 2602–2612.

[34]   Kotecha, J.H., and Djuric, P.M., "Gaussian Sum Particle Filtering for Dynamic State Space Models," *Proc. IEEE Int. Conf. Acoust. Speech, Signal Process*, 2001, pp. 3465–3468.

Appendix

Linearization Process for the EKF for Low-Cost Navigation

The focus of this book is to provide examples of MEMS-based inertial sensors for navigation of a land vehicle. This appendix gives the specific equations for GPS/MEMS INS integration for EKF implementation.

A.1 System Model for Loosely Coupled Approach

The state vector δx for an ENU frame are given below.

$$\delta x =$$

$$\left[\underbrace{\delta\varphi \quad \delta\lambda \quad \delta h}_{\delta r} \quad \underbrace{\delta V^e \quad \delta V^n \quad \delta V^u}_{\delta V} \quad \underbrace{\delta A^e \quad \delta A^n \quad \delta A^u}_{\delta A} \quad \underbrace{\delta\omega_x \quad \delta\omega_y \quad \delta\omega_z}_{\delta\omega=d} \quad \underbrace{\delta f_x \quad \delta f_y \quad \delta f_z}_{\delta f=b} \right]^T \quad (A.1)$$

The state vector for an NED frame is

$$\delta x =$$

$$\left[\underbrace{\delta\varphi \quad \delta\lambda \quad \delta h}_{\delta r} \quad \underbrace{\delta V^n \quad \delta V^e \quad \delta V^d}_{\delta V} \quad \underbrace{\delta A^n \quad \delta A^e \quad \delta A^d}_{\delta A} \quad \underbrace{\delta\omega_x \quad \delta\omega_y \quad \delta\omega_z}_{\delta\omega=d} \quad \underbrace{\delta f_x \quad \delta f_y \quad \delta f_z}_{\delta f=b} \right]^T \quad (A.2)$$

In (A.2), position errors, velocity, and attitude errors $(\delta r, \delta V, \delta A)$ are the navigation error states, and the other symbols are related to the sensor error models. The scale factors for gyros and accelerometers can also be added to (A.1) and (A.2). This will require the user to develop the appropriate relationship for the prediction and update stages.

A.1.1 Attitude Errors

Table A.1

Angular Rate	Derivative	Assumption
$\omega_{ie}^{\ell} = \begin{pmatrix} 0 \\ \omega^e \cos\varphi \\ \omega^e \sin\varphi \end{pmatrix}$	$\delta\omega_{ie}^{\ell} = \begin{pmatrix} 0 \\ -\delta\varphi\sin\varphi\omega^e \\ \delta\varphi\omega^e \cos\varphi \end{pmatrix}$	ω^e is constant and is given without any errors.
$\omega_{e\ell}^{\ell} = \begin{pmatrix} -\dfrac{V^n}{M+h} \\ \dfrac{V^e}{N+h} \\ \dfrac{V^e\tan\varphi}{N+h} \end{pmatrix}$	$\delta\omega_{e\ell}^{\ell} = \begin{pmatrix} -\dfrac{\delta V^n}{M+h} + \dfrac{\delta h V^n}{(M+h)^2} \\ \dfrac{\delta V^e}{N+h} - \dfrac{\delta h V^e}{(N+h)^2} \\ \dfrac{\delta V^e\tan\varphi}{N+h} + \dfrac{V^e\delta\varphi\sec^2\varphi}{N+h} - \dfrac{\delta h V^e\tan\varphi}{(N+h)^2} \end{pmatrix}$	Assume M and N are error free because the terms with $(M+h)^2$ and $(N+h)^2$ are very small.

The most extended form of the linearized attitude error in ENU frame is given below:

$$
\begin{bmatrix} \delta\dot{p} \\ \delta\dot{r} \\ \delta\dot{A} \end{bmatrix} = \begin{bmatrix} 0 & 1/(M+h) & 0 \\ -1/(N+h) & 0 & 0 \\ -\tan\varphi/(N+h) & 0 & 0 \end{bmatrix} \begin{bmatrix} \delta V^e \\ \delta V^n \\ \delta V^u \end{bmatrix} + R_b^l \begin{bmatrix} \delta\omega_x \\ \delta\omega_y \\ \delta\omega_z \end{bmatrix} +
$$

$$
\begin{bmatrix} 0 & 0 & -V^n/(M+h)^2 \\ \omega^e\sin\varphi & 0 & V^e/(N+h)^2 \\ -\omega^e\cos\varphi - V^e\sec^2\varphi/(N+h) & 0 & V^e\tan\varphi/(N+h)^2 \end{bmatrix} \begin{bmatrix} \delta\phi \\ \delta\lambda \\ \delta h \end{bmatrix} +
$$

$$
\begin{bmatrix} 0 & V^e\tan\varphi/(N+h)+\omega^e\sin\varphi & -V^e/(N+h)-\omega^e\cos\varphi \\ -V^e\tan\varphi/(N+h)-\omega^e\sin\varphi & 0 & -V^n/(M+h) \\ V^e/(N+h)+\omega^e\cos\varphi & V^n/(M+h) & 0 \end{bmatrix} \begin{bmatrix} \delta p \\ \delta r \\ \delta A \end{bmatrix}
$$

(A.3)

In the NED frame the attitude equation is:

$$\begin{bmatrix} \delta\dot{r} \\ \delta\dot{p} \\ \delta\dot{A} \end{bmatrix} = \begin{bmatrix} 0 & \dfrac{1}{N+h} & 0 \\ -\dfrac{1}{M+h} & 0 & 0 \\ 0 & -\dfrac{\tan\varphi}{N+h} & 0 \end{bmatrix} \begin{bmatrix} \delta V^n \\ \delta V^e \\ \delta V^d \end{bmatrix} + R^l_b \begin{bmatrix} \delta\omega_x \\ \delta\omega_y \\ \delta\omega_z \end{bmatrix} +$$

$$\begin{bmatrix} -\omega^e\sin\varphi & 0 & -\dfrac{V^e}{(N+h)^2} \\ 0 & 0 & \dfrac{V^n}{(M+h)^2} \\ -\omega^e\cos\varphi - \dfrac{V^e\sec^2\varphi}{(N+h)} & 0 & \dfrac{V^e\tan\varphi}{(N+h)^2} \end{bmatrix} \begin{bmatrix} \delta\phi \\ \delta\lambda \\ \delta h \end{bmatrix} + \qquad\text{(A.4)}$$

$$\begin{bmatrix} 0 & \dfrac{V^e\tan\varphi}{N+h} + \omega^e\sin\varphi & -\dfrac{V^n}{M+h} \\ -\dfrac{V^e\tan\varphi}{N+h} - \omega^e\sin\varphi & 0 & -\dfrac{V^e}{N+h} - \omega^e\cos\varphi \\ \dfrac{V^n}{M+h} & \dfrac{V^e}{N+h} + \omega^e\cos\varphi & 0 \end{bmatrix} \begin{bmatrix} \delta r \\ \delta p \\ \delta A \end{bmatrix}$$

A.1.2 Velocity Linearization

The linearized velocity error equation is:

$$\delta\dot{V}^\ell = \underbrace{\delta R^\ell_b f^b}_{term1} + \underbrace{R^\ell_b \delta f^b}_{term2} - \underbrace{(2\Omega^\ell_{ie} + \Omega^\ell_{e\ell})\delta V^\ell}_{term3} + \underbrace{(2\delta\Omega^\ell_{ie} + \delta\Omega^\ell_{e\ell})V^\ell}_{term4} + \underbrace{\delta g^\ell}_{term5} \quad\text{(A.5)}$$

The error of the gravity is given using free-air gravity reduction [1].

$$\delta g^\ell = \begin{bmatrix} 0 & 0 & -2\gamma / (R+h)\delta h \end{bmatrix}^T \qquad\text{(A.6)}$$

where γ is the normal gravity of the point of interest.

The most extended forms of the velocity errors are

$$
\begin{bmatrix} \delta\dot{v}^e \\ \delta\dot{v}^n \\ \delta\dot{v}^u \end{bmatrix} = \begin{bmatrix} 0 & f_u & -f_n \\ -f_u & 0 & f_e \\ f_n & -f_e & 0 \end{bmatrix} \begin{bmatrix} \delta p \\ \delta r \\ \delta A \end{bmatrix} + R_b^l \begin{bmatrix} \delta f_x \\ \delta f_y \\ \delta f_z \end{bmatrix} +
$$

$$
\begin{bmatrix} 2\omega^e \sin\varphi V^u + 2\omega^e \cos\varphi V^n + \dfrac{V^n V^e}{(N+h)\cos^2\varphi} - V^n V^e / [(N+h)\cos^2\varphi] & 0 & 0 \\[2ex] -2\omega^e \cos\varphi V^e - \dfrac{V^e V^e}{(N+h)\cos^2\varphi} & 0 & 0 \\[2ex] -2\omega^e \sin\varphi V^e & 0 & \dfrac{2\gamma}{R} \end{bmatrix} \begin{bmatrix} \delta\varphi \\ \delta\lambda \\ \delta h \end{bmatrix} + \quad (\text{A.7})
$$

$$
\begin{bmatrix} -\dfrac{V^u}{N+h} + \dfrac{\tan\varphi}{N+h} & 2\omega^e \sin\varphi + \dfrac{V^e \tan\varphi}{N+h} & -2\omega^e \cos\varphi - \dfrac{V^e}{N+h} \\[2ex] -2\omega^e \sin\varphi - \dfrac{2V^e \tan\varphi}{N+h} & -\dfrac{V^u}{M+h} & -\dfrac{V^n}{M+h} \\[2ex] 2\omega^e \cos\varphi + \dfrac{2V^e}{N+h} & \dfrac{2V^n}{M+h} & 0 \end{bmatrix} \begin{bmatrix} \delta V^e \\ \delta V^n \\ \delta V^u \end{bmatrix}
$$

$$
\begin{bmatrix} \delta\dot{v}^n \\ \delta\dot{v}^e \\ \delta\dot{v}^d \end{bmatrix} = \begin{bmatrix} 0 & f_d & -f_e \\ -f_d & 0 & f_n \\ f_e & -f_n & 0 \end{bmatrix} \begin{bmatrix} \delta r \\ \delta p \\ \delta A \end{bmatrix} + R_b^l \begin{bmatrix} \delta f_x \\ \delta f_y \\ \delta f_z \end{bmatrix} +
$$

$$
\begin{bmatrix} 2V^e \omega^e \cos\varphi - \dfrac{(V^e)^2}{(N+h)\cos^2\varphi} & 0 & \dfrac{-V^n V^d}{(M+h)^2} + \dfrac{(V^e)^2 \tan\varphi}{(N+h)^2} \\[2ex] 2\omega^e (V^n \cos\varphi - V^d \sin\varphi) + \dfrac{V^e V^n}{(N+h)\cos^2\varphi} & 0 & \dfrac{-V^e V^d}{(N+h)^2} - \dfrac{V^n V^e \tan\varphi}{(N+h)^2} \\[2ex] 2\omega^e \sin\varphi V^e & 0 & \dfrac{(V^e)^2}{(N+h)^2} + \dfrac{(V^n)^2 \tan\varphi}{(M+h)^2} - \dfrac{2\gamma}{(R+h)} \end{bmatrix} \begin{bmatrix} \delta\varphi \\ \delta\lambda \\ \delta h \end{bmatrix} + \quad (\text{A.8})
$$

$$
\begin{bmatrix} \dfrac{V^d}{M+h} & -2\omega^e \sin\varphi - 2\dfrac{V^e \tan\varphi}{N+h} & \dfrac{V^n}{M+h} \\[2ex] 2\omega^e \sin\varphi + \dfrac{V^e \tan\varphi}{N+h} & \dfrac{V^d + V^n \tan\varphi}{N+h} & 2\omega^e \cos\varphi + \dfrac{V^e}{N+h} \\[2ex] -\dfrac{2V^n}{M+h} & -2\omega^e \cos\varphi - \dfrac{2V^e}{N+h} & 0 \end{bmatrix} \begin{bmatrix} \delta V^n \\ \delta V^e \\ \delta V^d \end{bmatrix}
$$

A.1.3 Position Linearization

$$\delta \dot{r}^{\ell} = \begin{pmatrix} \delta \dot{\varphi} \\ \delta \dot{\lambda} \\ \delta \dot{h} \end{pmatrix} = \begin{pmatrix} 0 & 1/(M+h) & 0 \\ 1/(N+h)\cos\varphi & 0 & 0 \\ 0 & 0 & 1 \end{pmatrix} \begin{pmatrix} \delta V^{e} \\ \delta V^{n} \\ \delta V^{u} \end{pmatrix} +$$

$$\begin{pmatrix} 0 & 0 & -\dfrac{V^{n}}{M+h} \\ V^{e}\tan\varphi \,/\,(N+h)\cos\varphi & 0 & -V^{e}\,/\,(N+h)^{2}\cos\varphi \\ 0 & 0 & 0 \end{pmatrix} \begin{bmatrix} \delta\varphi \\ \delta\lambda \\ \delta h \end{bmatrix} \qquad \text{(A.9)}$$

The position equation in the NED frame is given in (A.10).

$$\delta \dot{r}^{\ell} = \begin{pmatrix} \delta \dot{\varphi} \\ \delta \dot{\lambda} \\ \delta \dot{h} \end{pmatrix} = \begin{pmatrix} \dfrac{1}{M+h} & 0 & 0 \\ 0 & \dfrac{1}{(N+h)\cos\varphi} & 0 \\ 0 & 0 & -1 \end{pmatrix} \begin{pmatrix} \delta V^{n} \\ \delta V^{e} \\ \delta V^{d} \end{pmatrix} +$$

$$\begin{pmatrix} 0 & 0 & -\dfrac{V^{n}}{\left(M+h\right)} \\ \dfrac{V^{e}\sin\varphi}{(N+h)\cos^{2}\varphi} & 0 & -\dfrac{V^{e}}{(N+h)^{2}\cos\varphi} \\ 0 & 0 & 0 \end{pmatrix} \begin{bmatrix} \delta\varphi \\ \delta\lambda \\ \delta h \end{bmatrix} \qquad \text{(A.10)}$$

A.1.4 Sensor Errors

After the removal of the deterministic parts of the inertial sensor biases, it is a common practice to model the residual stochastic part as a first-order Gauss-Markov process [2].

A.2 GPS Measurement Model

In the loosely coupled approach, position and velocity data from GPS is used to make the measurement matrix as follows:

$$\delta \tilde{z}_k = \begin{bmatrix} \varphi_{DGPS} - \varphi_{INS} & \lambda_{DGPS} - \lambda_{INS} & h_{DGPS} - h_{INS} \end{bmatrix}^T \quad (A.11)$$

The corresponding measurement matrix H_k will be

$$H_k = \begin{bmatrix} 1 & 0 & 0 & 0 & 0 & 0 & 0 & 0 & 0 & 0 & 0 & 0 & 0 & 0 & 0 \\ 0 & 1 & 0 & 0 & 0 & 0 & 0 & 0 & 0 & 0 & 0 & 0 & 0 & 0 & 0 \\ 0 & 0 & 1 & 0 & 0 & 0 & 0 & 0 & 0 & 0 & 0 & 0 & 0 & 0 & 0 \end{bmatrix} \quad (A.12)$$

The measurement noise matrix obtained from GPS has the following form:

$$R_k = \begin{bmatrix} \sigma_\varphi^2 & 0 & 0 \\ 0 & \sigma_\lambda^2 & 0 \\ 0 & 0 & \sigma_h^2 \end{bmatrix} \quad (A.13)$$

The measurement equations and measurement matrix is the same whether an NED or ENU l-frame implementation is chosen for the navigation Kalman filter.

A.3 System Model for the Tightly Coupled Approach

The error state vector for tightly coupled integration consists of the navigation states, sensor error states, and clock-related error states as shown in Table A.2 in detail.

Table A.2

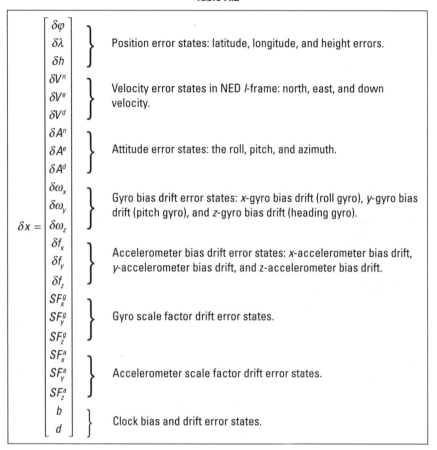

$$\delta x = \begin{bmatrix} \delta\varphi \\ \delta\lambda \\ \delta h \\ \delta V^n \\ \delta V^e \\ \delta V^d \\ \delta A^n \\ \delta A^e \\ \delta A^d \\ \delta\omega_x \\ \delta\omega_y \\ \delta\omega_z \\ \delta f_x \\ \delta f_y \\ \delta f_z \\ SF^g_x \\ SF^g_y \\ SF^g_z \\ SF^a_x \\ SF^a_y \\ SF^a_z \\ b \\ d \end{bmatrix}$$

Position error states: latitude, longitude, and height errors.

Velocity error states in NED *l*-frame: north, east, and down velocity.

Attitude error states: the roll, pitch, and azimuth.

Gyro bias drift error states: *x*-gyro bias drift (roll gyro), *y*-gyro bias drift (pitch gyro), and *z*-gyro bias drift (heading gyro).

Accelerometer bias drift error states: *x*-accelerometer bias drift, *y*-accelerometer bias drift, and z-accelerometer bias drift.

Gyro scale factor drift error states.

Accelerometer scale factor drift error states.

Clock bias and drift error states.

The psi-angle error model [3] is provided. Theoretically, there is no difference between the *c*-frame and the *l*-frame; but, the two frames are misaligned due to the Earth's rotation and transport rate in the absence of updates. This misalignment is caused solely because of the drift errors in gyros.

This misalignment is written as

$$\delta\theta = [\delta\lambda\cos\varphi \quad -\delta\varphi \quad -\delta\lambda\sin\varphi]^T \qquad (A.14)$$
$$= [\delta r_E/(N+h) \quad \delta r_N/(M+h) \quad -\delta r_E\tan\varphi/(N+h)]^T$$

The two frames, the *l*-frame and the *c*-frame, are shown in (A.15) and (A.16):

$$\hat{v}^l = v^l + \delta v^l = v^c + \delta v^c \tag{A.15}$$

$$\hat{g}^l = g^l + \delta g^l = g^c + \delta g^c \tag{A.16}$$

where $\delta v^l = \delta v^c - \delta\theta \times v^c$ and $\delta g^l = \delta g^c - \delta\theta \times g^c$.

Using these two values, the value of δg^c that is used in the mechanization equation in the c-frame is estimated as:

$$\delta g^l = \begin{bmatrix} 0 & 0 & 2g\delta r_D /(R + h) \end{bmatrix}^T \tag{A.17}$$

$$\delta\theta \times g^c = \begin{bmatrix} -g\delta r_N /(M + h) & -g\delta r_E /(N + h) & 0 \end{bmatrix}^T \tag{A.18}$$

The Earth's rotation rate error is given by (A.19).

$$\delta\omega_{ie}^l = -\delta\theta \times \omega_{ie}^c \tag{A.19}$$

After considering the error properties and the above relationships, simplified mechanization equations are produced for the c-frame implementation.

$$\delta\dot{r}^c = -\omega_{ec}^c \times \delta r^c + \delta v^c$$

$$\delta\dot{v}^c = f^c \times \Psi - \left(2\omega_{ie}^c + \omega_{ec}^c\right) \times \delta v^c + \delta g^c + R_b^c \delta f^b \tag{A.20}$$

$$\dot{\Psi} = -\left(\omega_{ie}^c + \omega_{ec}^c\right) \times \Psi - R_b^c \delta\omega_{ib}^b$$

The c-frame attitude is the only different entity that differentiates between the c- and l-frame and to add to the error states, these attitudes are converted into the usual roll, pitch, and azimuth representation. The conversion is simply taking the misalignments into account for the two frames as discussed before.

The position error dynamics have two error components given by (A.20). The first error is actually the error in position due to the rotation rate A_{pos}^{pos} with time and the second part is due to the velocity errors A_{pos}^{vel}, which is estimated by simple time integration.

$$A_{pos}^{pos} = \begin{bmatrix} 1 & \omega_{el}^l(3,1)dt & -\omega_{el}^l(2,1)dt \\ -\omega_{el}^l(3,1)dt & 1 & \omega_{el}^l(1,1)dt \\ \omega_{el}^l(2,1)dt & -\omega_{el}^l(1,1)dt & 1 \end{bmatrix} \tag{A.21}$$

$$A_{pos}^{vel} = I_{3x3}dt \tag{A.22}$$

where $I_{3x3} = \begin{bmatrix} 1 & 0 & 0 \\ 0 & 1 & 0 \\ 0 & 0 & 1 \end{bmatrix}$.

Let us start by looking at the different components of the velocity error dynamic equation (A.20). The first part is related to the position error A_{vel}^{pos}, which occurs due to gravity errors.

$$A_{vel}^{pos} = \begin{bmatrix} -g \ / \ (M + h)dt & 0 & 0 \\ 0 & -g \ / \ (N + h)dt & 0 \\ 0 & 0 & 2g \ / \ (R + h)dt \end{bmatrix} \tag{A.23}$$

Furthermore, the transition matrix term for velocity due to IMU velocity is related to the angular rates as shown in the velocity mechanization equation. Additionally, the transition matrix also models the attitude and accelerometer biases.

$$A_{vel}^{vel} = \begin{bmatrix} 1 & \hat{\omega}(3,1)dt & -\hat{\omega}(2,1)dt \\ -\hat{\omega}(3,1)dt & 1 & \hat{\omega}(1,1)dt \\ \hat{\omega}(2,1)dt & -\hat{\omega}(1,1)dt & 1 \end{bmatrix} \tag{A.24}$$

where $\hat{\omega} = -2\omega_{ie}^e - \omega_{ec}^e = -2\omega_{ie}^l - \omega_{el}^l$.

$$A_{vel}^{att} = \begin{bmatrix} 0 & -f^l(3,1)dt & f^l(2,1)dt \\ f^l(3,1)dt & 0 & -f^l(1,1)dt \\ -f^l(2,1)dt & f^l(1,1)dt & 0 \end{bmatrix} \tag{A.25}$$

$$A_{vel}^{accel\_bias} = R_b^l dt \tag{A.26}$$

$$A_{vel}^{accel\_SF} = R_b^l diag(f_x^b, f_y^b, f_z^b)dt \tag{A.27}$$

The attitude errors components are also given in (A.28).

$$A_{att}^{att} = \begin{bmatrix} 1 & \omega_{ic}^c(3,1)dt & -\omega_{ic}^c(2,1)dt \\ -\omega_{ic}^c(3,1)dt & 1 & \omega_{ic}^c(1,1)dt \\ \omega_{ic}^c(2,1)dt & -\omega_{ic}^c(1,1)dt & 1 \end{bmatrix} \qquad \text{(A.28)}$$

where $\omega_{ic}^c = \omega_{il}^l = \omega_{ie}^l + \omega_{el}^l$.

$$A_{att}^{gyro\_bias} = -R_b^l dt \qquad \text{(A.29)}$$

$$A_{att}^{gyro\_SF} = -R_b^l diag(\omega_{ib}^b(1,1), \omega_{ib}^b(2,1), \omega_{ib}^b(3,1))dt \qquad \text{(A.30)}$$

Inertial sensor errors are generally modeled as first-order Gauss-Markov processes [2].

$$A_{gyro\_bias}^{gyro\_bias} = diag(-\beta_x^b, -\beta_y^b, -\beta_z^b) \qquad \text{(A.31)}$$

$$A_{acc\_bias}^{acc\_bias} = diag(-\alpha_x^b, -\alpha_y^b, -\alpha_z^b) \qquad \text{(A.32)}$$

$$A_{gyro\_SF}^{gyro\_SF} = diag(-\beta_x^{SF}, -\beta_y^{SF}, -\beta_z^{SF} \qquad \text{(A.33)}$$

$$A_{acc\_SF}^{acc\_SF} = diag(-\alpha_x^{SF}, -\alpha_y^{SF}, -\alpha_z^{SF}) \qquad \text{(A.34)}$$

where α, and β are time constants of the first-order Gauss-Markov process [2].

GPS receiver clock error states are unique to tightly coupled integration schemes [4–12]. The clock bias can be written as a random walk process as follows:

$$b_t = b_{t-1} + (d_t)dt \qquad \text{(A.35)}$$

where d is the random constant error for the clock drift for each time step:

$$d_t = d_{t-1} \qquad \text{(A.36)}$$

Therefore, the transition matrix for the clock bias and drift errors is written as

$$A_{bd}^{bd} = \begin{bmatrix} 1 & dt \\ 0 & 1 \end{bmatrix} \qquad \text{(A.37)}$$

Combining (A.35)–(A.37) gives the transition matrix:

$$
\Phi = \begin{bmatrix}
A_{pos}^{pos} & A_{pos}^{vel} & 0_{3\times3} & 0_{3\times3} & 0_{3\times3} & 0_{3\times3} & 0_{3\times3} & 0_{1\times2} \\
A_{vel}^{pos} & A_{vel}^{vel} & A_{vel}^{att} & 0_{3\times3} & A_{vel}^{acc\_bias} & 0_{3\times3} & A_{vel}^{acc\_SF} & 0_{1\times2} \\
0_{3\times3} & 0_{3\times3} & A_{att}^{att} & A_{att}^{gyro\_bias} & 0_{3\times3} & A_{att}^{gyro\_SF} & 0_{3\times3} & 0_{1\times2} \\
0_{3\times3} & 0_{3\times3} & 0_{3\times3} & A_{gyro\_bias}^{gyro\_bias} & 0_{3\times3} & 0_{3\times3} & 0_{3\times3} & 0_{1\times2} \\
0_{3\times3} & 0_{3\times3} & 0_{3\times3} & 0_{3\times3} & A_{acc\_bias}^{acc\_bias} & 0_{3\times3} & 0_{3\times3} & 0_{1\times2} \\
0_{3\times3} & 0_{3\times3} & 0_{3\times3} & 0_{3\times3} & 0_{3\times3} & A_{gyro\_SF}^{gyro\_SF} & 0_{3\times3} & 0_{1\times2} \\
0_{3\times3} & 0_{3\times3} & 0_{3\times3} & 0_{3\times3} & 0_{3\times3} & 0_{3\times3} & A_{acc\_SF}^{acc\_SF} & 0_{1\times2} \\
0_{2\times3} & 0_{2\times3} & 0_{2\times3} & 0_{2\times3} & 0_{2\times3} & 0_{2\times3} & 0_{2\times3} & A_{bd}^{bd}
\end{bmatrix}
$$

$$(A.38)$$

A.4 The Update Stage

The update stage for tightly coupled integration is different from the loosely coupled integration strategy. Here, raw measurements (i.e., pseudorange and Doppler measurements), which are sufficient for low-cost MEMS-based integrated navigation, are used. The first step is to convert the raw measurements into required observables for the integration filter. The pseudorange is the distance between the satellite and the GPS receiver with errors associated with the transportation medium and the clock errors.

$$\tilde{\rho} = \rho + cdt_{clk,sat} - cdt_{clk} + d_{Iono} + d_{Tropo} + d_{multipath} + \varepsilon_{\rho} \quad (A.39)$$

where $\tilde{\rho}$ is the measured pseudorange, ρ is the true pseudorange, c is the speed of light (2.9979×10^8 m/s), dt_{clk} is the error due to receiver clock bias, $dt_{clk,sat}$ is the error due to the satellite clock, d_{Iono} is the error due to the ionosphere, d_{Tropo} is the error due to the troposphere, $d_{multipath}$ is the error due to the multipath, and ε_p is the noise.

The relative motion of the satellite and user results in changes in the observed frequency and is commonly known as a Doppler shift or range rate. If the satellite velocity is known and the Doppler shift is measured, the user velocity can be estimated. The GPS receiver measures the Doppler shift in the carrier tracking loop and it is readily available as one of the observations. The range rate can be given as follows:

$$\tilde{\dot{\rho}} = \dot{\rho} + d\dot{\rho} + f_{L1}(d\dot{t}_{clk,sat} - d\dot{t}_{clk}) - \dot{d}_{Iono} + \dot{d}_{Tropo} + \varepsilon \qquad \text{(A.40)}$$

where $\tilde{\dot{\rho}}$ is the measured Doppler, $\dot{\rho}$ is the Doppler or range rate, $d\dot{\rho}$ is the orbital errors drift, f_{L1} is the L1 carrier frequency, $d\dot{t}_{clk,sat}$ is the satellite clock drift, $d\dot{t}_{clk}$ is the receiver clock drift, $\dot{d}_{Iono}$ is the ionospheric delay drift, $\dot{d}_{Tropo}$ is the tropospheric delay drift, and ε is the noise.

A.5 Pseudorange and Doppler Corrections

The measurements for the tightly coupled integration consist of pseudorange (A.39) and Doppler (A.40) for each available satellite. Both equations consist of the required parameter along with a number of error terms. After calculating and removing all the error terms, the corrected measurements are then used in the measurement vector $\delta\tilde{z}_k$ of the KF. If there are n satellites available at the kth epoch, the measurement vector is given as:

$$\delta\tilde{z}_k = \begin{bmatrix} \rho_{INS,1}^k - \tilde{\rho}_{GPS,1}^k \\ \vdots \\ \rho_{INS,n}^k - \tilde{\rho}_{GPS,n}^k \\ \dot{\rho}_{INS,1}^k - \tilde{\dot{\rho}}_{GPS,1}^k \\ \vdots \\ \dot{\rho}_{INS,n}^k - \tilde{\dot{\rho}}_{GPS,n}^k \end{bmatrix} \qquad \text{(A.41)}$$

where $\rho_{INS,j}^k$ is the computed pseudorange using INS measurements for the jth satellite, $\dot{\rho}_{INS,j}^k$ is the INS computed range rate for the satellite, $\tilde{\rho}_{INS,j}^k$ is the corrected pseudorange for the satellite, and $\tilde{\dot{\rho}}_{INS,j}^k$ is the Doppler for the satellite.

References

[1]   Featherstone, W.E., and Dentith, M.C., "A Geodetic Approach to Gravity Data Reduction for Geophysics," Vol. 23, No. 10, December 10, 1997, pp. 1063–1070.

[2]   Maybeck, P.S., *Stochastic Models, Estimation and Control,* Volume 3, New York: Academic Press Inc., 1982.

[3]   Savage, P., *Strapdown Analytics*, Part 1 & 2, MN: Strapdown Associates, 2000.

[4]   Godha, S. "Performance Evaluation of Low Cost MEMS-Based IMU Integrated with GPS for Land Vehicle Navigation Application," M.Sc. thesis, Department of Geomatics Engineering, University of Calgary, Canada, UCGE Report No. 20239, 2006, available online: http://www.ucalgary.ca/engo_webdocs/MEC/06.20239.SGodha.pdf.

[5]   Grewal, M.S., Weill, L., and Andrews, A.P., *Global Positioning Systems, Inertial Navigation, and Integration, Second Edition*, New Jersey: John Wiley & Sons, 2007.

[6]   Hofmann-Wellenhof, B., Lichtenegger, H., and Collins, J., *GPS Theory and Practice, Fifth Edition*, Austria: Springer, 2004.

[7]   ICD-GPS-200C, "GPS Interface Control Document, NAVSTAR GPS Space Segment / Navigation User Interfaces," IRN-200C-004, 2000, http://www.navcen.uscg.gov/gps.

[8]   Kaplan, E.D., *Understanding GPS: Principles and Applications*, Norwood, MA: Artech House, 1996.

[9]   Knight, D.T., "Rapid Development of Tightly Coupled GPS/INS Systems," *Aerospace and Electronic Systems Magazine*, Vol. 12, No. 2, 1999, pp. 14–18.

[10]   Misra, P., and Enge, P., *Global Positioning System: Signals, Measurements, and Performance*, Lincoln, MA: Ganga-Jamuna Press, 2001.

[11]   Parkinson, B.W., and Spilker Jr., J.J., "Global Positioning System: Theory and Applications," in *Progress in Astonautics and Aeronautics*, United States: AIAA, 1996.

[12]   Tsui, J.B., *Fundamentals of Global Positioning System Receivers: A Software Approach, Second Edition*, Canada: John Wiley & Sons, Inc., 2005.

About the Authors

Priyanka Aggarwal is a research consultant at the University of Michigan. She obtained her Ph.D. in geomatics engineering in 2009 from the University of Calgary. Her master's degree is in electrical and computer engineering (2004) from the University of Calgary, and her bachelor's degree in electronics and communication (2000) is from the University of Kurukshetra, India.

Her research interests are integrated navigation and positioning techniques for ground and airborne applications. Her expertise includes developing algorithms for integrated inertial navigation using Kalman, unscented, and particle filters; real-time kinematic positioning using map-matching techniques; and autonomous pedestrian navigation systems in GPS-denied atmospheres. She has written several high-quality journal papers, holds a patent on hybrid particle filtering, and has contributed to 25 research publications.

Zainab Syed obtained her B.Sc. in geomatics engineering in 2001, her M.Sc. in electrical and computers engineering in 2004, and her Ph.D. in geomatics engineering in 2009 from the University of Calgary. Her current research focuses on very low-cost and portable navigation systems. Dr. Syed is a cofounder and VP of technology at the Trusted Positioning Inc. Over the course of her career, Dr. Syed had been involved in many collaborative R&D with industrial partners such as SiRF Inc. and NovAtel Inc. To date, Dr. Syed has over 30 publications, two patents pending, and software for integrated navigation systems. The earlier part of her career was focused on the optimization of data processing techniques for highly nonlinear systems using parallel processes.

Aboelmagd Noureldin is a cross-appointment associate professor at the Departments of Electrical and Computer Engineering of both Queen's University and the Royal Military College (RMC) of Canada. He is also the founder and leader of the Navigation and Instrumentation research group at RMC. His research is related to artificial intelligence, digital signal processing, spectral estimation and denoising, wavelet multiresolution analysis, nonlinear estimation, and adaptive filtering with emphasis on their applications in mobile multisensor system integration for navigation and positioning technologies.

He is the developer of several innovative methods for reliable navigation in urban and indoor areas as well as other denied or challenging GPS environments. Dr. Noureldin holds a B.Sc. in electrical engineering (1993) and an M.Sc. in engineering physics (1997) from Cairo University, Giza, Egypt. In addition, he holds a Ph.D. in electrical and computer engineering (2002) from the University of Calgary, Alberta, Canada. Dr. Noureldin is a senior member of the IEEE.

Naser El-Sheimy is a professor of the Department of Geomatics Engineering, Schulich School of Engineering (SSE), at the University of Calgary and is also the scientific director at the Integrated Resource Management (IRM) Centre of Excellence for Commercialization of Research (CECR). He holds a Canada research chair (CRC) in mobile multisensor systems. His research interests include multisensor systems, mobile mapping systems, estimation techniques, real-time kinematic positioning, and digital photogrammetric along with their applications in transportation, mapping, and geospatial information systems (GIS).

Dr. El-Sheimy has published a book and over 300 papers in academic journals, conferences, and workshop proceedings, in which he has received over 20 national and international paper awards. He organized and participated in organizing many national and international conferences. His research on multisensor systems applications has been recognized internationally.

Index

Ubiquitous Positioning, Robin Mannings

Wireless Positioning Technologies and Applications, Alan Bensky

For further information on these and other Artech House titles, including previously considered out-of-print books now available through our In-Print- Forever® (IPF®) program, contact:

Artech House Publishers
685 Canton Street
Norwood, MA 02062
Phone: 781-769-9750
Fax: 781-769-6334
e-mail: artech@artechhouse.com

Artech House Books
16 Sussex Street
London SW1V 4RW UK
Phone: +44 (0)20 7596 8750
Fax: +44 (0)20 7630 0166
e-mail:
artech-uk@artechhouse.com

Find us on the World Wide Web at: www.artechhouse.com